Bo Nilsson
Exercises with Solutions in Radiation Physics

Bo Nilsson

Exercises with Solutions in Radiation Physics

Managing Editor: Paulina Leśna-Szreter

Language Editor: Andrew Laister

DE GRUYTER
OPEN

Published by De Gruyter Open Ltd, Warsaw/Berlin
Part of Walter de Gruyter GmbH, Berlin/Munich/Boston

Copyright © 2015 Bo Nilsson
published by De Gruyter Open

ISBN 978-3-11-044205-2
e-ISBN 978-3-11-044206-9

Bibliographic information published by the Deutsche Nationalbibliothek
The Deutsche Nationalbibliothek lists this publication in the Deutsche Nationalbibliografie;
detailed bibliographic data are available in the Internet at http://dnb.dnb.de.

www.degruyteropen.com

Cover illustration: © iStock

Contents

Preface

This material is intended for use in courses in radiation physics. Many textbooks include exercises, but not often full solutions, and they often refer to specific material in the textbook. This material can be used in many courses, often included in a medical physics graduate program, independent of a specific textbook. The material consists of six chapters covering the basic fundamental radiation physics, but not the more specific clinical applications where there is a rapid change and in which the exercises may be obsolete after some time. The first chapter includes exercises related to radioactive sources and decay schemes. This is followed by a chapter covering the interaction of ionizing radiation, including photons and charged particles. The text then continues with a chapter on detectors and measurements including both some simple counting statistics and properties of detectors. The next chapter is dedicated to dosimetry, which is a major subject in medical physics. A short chapter is covering radiobiology, where there is a focus on different cell survival models. The last chapter is dealing with radiation protection and health physics. Both radiation shielding calculations and radioecology are covered. The exercises in the material have been used in the education for medical physicists in Stockholm, Sweden, and the order of chapters follows the order of courses in this education, but hopefully they are useful in all applications of radiation physics including also health physics. Some problems are probably similar to what can be found in other material as there are some items that are always important to include in the courses and when producing exercises it is easy to forget where any idea is coming from.

The student is supposed to have a background in mathematics and physics corresponding to a BSc in physics. The mathematics involved is mainly straightforward and includes only basic integrals and differential equations. Each chapter starts with a small refreshment of important definitions and relations that are useful for the chapter. The material is not aimed to be a textbook and for a deeper knowledge and understanding the reader is referred to ordinary textbooks, some of which are listed in the bibliography. The chapter then continues with a section of exercises followed by a section with solutions. The reader is recommended to make an effort to understand the exercise and try to solve it before checking the proposed solution. Most exercises are numerical and a numerical answer is expected. Some exercises are more intended for a discussion where probably different answers can be acceptable. In order to solve the exercise and obtain a numerical value, data like decay constants, interaction coefficients etc are often needed. They are normally not included in the exercise, but the reader is expected to understand which data are needed and then find them in relevant tabulated material. Much of the information is obtained at different sites on the Internet, and some sites are listed in the bibliography. However, some data may be difficult to find, and for those are tables included in this material. The reference list at the end of the book is divided in two sections, one including tables and sites that could be useful when solving the exercises and one including some typical textbooks for the different courses related to the different chapters. This list is far from complete, but gives just some examples of books used by education establishments in Stockholm. There is also after some chapters, specific references to papers mentioned in the chapter.

Many of these exercises have been used during several years in the education for hospital physics in Stockholm. However, in spite of this, there are probably both typing mistakes or badly explained solutions. I am of course grateful if these mistakes are reported to me.

Finally, I would like to acknowledge help from my colleagues at the department, Prof. Irena Gudowska, ass Prof, Albert Siegbahn, and ass Prof. Iuliana Toma-Dasu; who have encouraged me to make this collection of exercises and also have pointed out several mistakes in the material. Without their contribution the mistakes would be even more abundant. I also want to thank all students that have been working with the problems during several years.

Stockholm, June 2014

Bo Nilsson

1 Radiation Sources and Radioactive Decay

1.1 Definitions and Equations

1.1.1 Radioactivity and Decay Equations

Activity
Activity is defined as

$$A = \frac{dN}{dt} = \lambda N \quad \text{Unit}: 1\,\text{Bq (becquerel)} = 1\,\text{s}^{-1} \tag{1.1.1}$$

where A is activity, dN/dt is the number of spontaneous nuclear transformations, dN, from a particular energy state in a time interval dt. λ is the decay constant (s^{-1}) and N is the number of radioactive nuclei.

The specific activity is defined as the activity of a certain radionuclide per mass unit (Bq kg^{-1}).

$$C = \frac{A}{m} \tag{1.1.2}$$

Radioactive decay
A radionuclide decays according to the equation

$$N_1(t) = N_0 e^{-\lambda t} = N_0 e^{-t \ln 2/T} \tag{1.1.3}$$

where $N_1(t)$ is the number of radioactive nuclides after a time t, $N_0 = N_1(0)$ is the number of radioactive nuclides at time 0 and T is the half-life ($T = \ln 2/\lambda$).

The equation may also be expressed as

$$A_1(t) = A_0 e^{-\lambda t} = A_0 e^{-t \ln 2/T} \tag{1.1.4}$$

Sometimes the daughter nuclides are also radioactive and a chain of radioactive nuclides is obtained. A general solution for the activity of a radionuclide in the chain is given by the Bateman equations. In this compilation only the first three radionuclides in the chain will be treated.

$$N_1(t) = N_0 e^{-\lambda_1 t} \tag{1.1.5}$$

$$N_2(t) = N_0 \frac{\lambda_1}{\lambda_2 - \lambda_1}(e^{-\lambda_1 t} - e^{-\lambda_2 t}) \tag{1.1.6}$$

$$N_3(t) = N_0 \lambda_1 \lambda_2 \left[\frac{e^{-\lambda_1 t}}{(\lambda_3 - \lambda_1)(\lambda_2 - \lambda_1)} + \frac{e^{-\lambda_2 t}}{(\lambda_3 - \lambda_2)(\lambda_1 - \lambda_2)} + \frac{e^{-\lambda_3 t}}{(\lambda_2 - \lambda_3)(\lambda_1 - \lambda_3)}\right] \tag{1.1.7}$$

All these equations assume that $N_2(0)$ and $N_3(0)$ are equal to zero. If not, corrections have to be made, by adding to the equation above, the activity of the separate radionuclides at $t = 0$ corrected for the decay to the time t. E.g.

$$N_2'(t) = N_2'(0)e^{-\lambda_2 t} \tag{1.1.8}$$

where $N_2'(0)$ is the number of N_2' radionuclides at time, $t=0$. In some situations Eq. (1.1.6) may be simplified, as shown below.

If $\lambda_1 << \lambda_2$, then

$$N_2(t) = N_0 \frac{\lambda_1}{\lambda_2}(1 - e^{-\lambda_2 t}) \tag{1.1.9}$$

For large values of t, this equation is reduced to (secular equilibrium)

$$\lambda_1 N_1 = \lambda_2 N_2 \tag{1.1.10}$$

If there is a chain of radionuclides with a long lived parent radionuclide, then this simplification may be extended to the whole chain.

$$\lambda_1 N_1 = \lambda_2 N_2 = \lambda_3 N_3 = \ldots = \lambda_k N_k = \ldots \tag{1.1.11}$$

If $\lambda_1 < \lambda_2$ the equation for large t is reduced to (transient equilibrium)

$$N_2(t) = N_0 \frac{\lambda_1}{\lambda_1 - \lambda_2} e^{-\lambda_1 t} \tag{1.1.12}$$

Production of radionuclides

The number of produced radionuclides at time t is given by the equation

$$N(t) = \frac{\sigma \dot{\Phi} m N_A}{m_a(\lambda - \sigma \dot{\Phi})}(e^{-\sigma \dot{\Phi} t} - e^{-\lambda t}) \tag{1.1.13}$$

where σ is the activation cross section (m^2), $\dot{\Phi}$ is the particle fluence rate ($m^{-2}\ s^{-1}$), $N_A = 6.022 \cdot 10^{26}$ ($kmol^{-1}$) is Avogadro's number, m_a is the atomic mass of the target atom, and m is the mass of the target (kg).

If the decrease of the number of target nuclei during activation can be neglected the equation is reduced to

$$N(t) = \frac{\sigma \dot{\Phi} m N_A}{m_a \lambda}(1 - e^{-\lambda t}) \tag{1.1.14}$$

The corresponding equation for the activity is

$$A(t) = \frac{\sigma \dot{\Phi} m N_A}{m_a}(1 - e^{-\lambda t}) \tag{1.1.15}$$

1.1.2 Disintegration Schematics

In many situations it is important to know the number of particles emitted per decay and their energies, e.g. when determining the activity of a sample using spectroscopic measurements, or in internal dosimetry. In particular when there are nuclei with excited levels, the de-excitation can lead to both photons and electrons of different energies with different probabilities.

A full calculation including all particles is often complicated and in practice one should use tables such as the Table of Radioactive Isotopes (Browne and Firestone, 1986), MIRD: Radionuclide Data and Decay schemes (Eckerman and Endo, 2008), the Table of Radioactive Isotopes (ICRP, 1986) or Nuclear Data Tables (ICRP 2001), but for illustration a simplified example of the decay of ^{137}Cs is discussed here (Fig 1.1). ^{137}Cs is a β-emitting radionuclide. As ^{137}Cs has a half life of 30 y and ^{137m}Ba a half life of 2.55 min, it is often assumed that ^{137}Cs and ^{137m}Ba are in radioactive equilibrium.

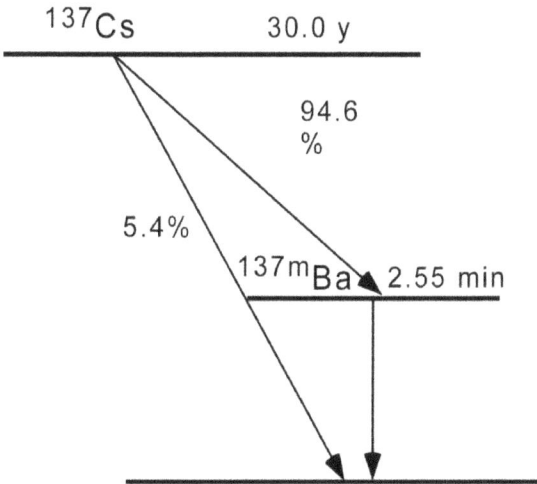

Figure 1.1: Decay scheme for ^{137}Cs.

I. Decay of ^{137}Cs

^{137}Cs emits two β-particles with different energies and frequencies. 5.4% of the decays go to the stable level of ^{137}Ba, and 94.6% go to the metastable level ^{137m}Ba. The β-particles are emitted giving a spectrum of energies, as part of the decay energy goes to the neutrino, and both the maximum energy and the mean energy are of interest. The maximum energy is obtained from the energy released in the decay, Q_β, which is obtained from the difference in mass of the parent nucleus and the daughter nucleus. As the β-particle has a mass that is very small compared to the parent nucleus, the total decay energy can be assumed to be distributed between the β-particle and the neutrino. The Q_β-value of the ^{137}Cs-decay is 1.175 MeV. Thus the maximum energy for β_1 resulting in the daughter nucleus in a stable state is 1.175 MeV. The maximum energy of β_2, going to the excited state of ^{137m}Ba is obtained by subtracting the energy of the excited level, 0.6616 MeV giving $1.175 - 0.6616 = 0.513$ MeV.

The shape of the β-particle spectrum is dependent on the type of the decay, the decay energy and the atomic number (See Table 1.1). Data for the mean energy for the different types of decay may be found e.g. in MIRD (Eckerman and Endo, 2008).

Table 1.1: Frequency, mean and maximum energies of the β-particles emitted in the ^{137}Cs-decay.

Particle	Frequency	Mean energy Mev	Maximum energy Mev	Type of decay
β_1	0.054	$1.175 \cdot 0.361 = 0.423$	1.175	2nd forbidden
β_2	0.946	$0.513 \cdot 0.34 = 0.175$	0.513	Unique 1st forbidden

II. De-excitation of ^{137m}Ba

The excitation energy can be released either by emission of a γ-photon or an internal conversion electron. In the internal conversion electron process, the excitation energy is transferred to an electron in an electron shell, mainly the K-shell. The probability that an electron will be emitted is given by the internal conversion coefficient α defined as $\alpha = N_e / N_\gamma$. $\alpha_K = N_{eK} / N_\gamma$ gives only the internal conversion electrons from the K-shell and so on. N_e is the number of internal conversion electrons per de-excitation and N_γ is the number of γ-photons per de-excitation.

The electron will obtain the excitation energy minus the binding energy of the electron. The emission of internal conversion electrons will result in vacancies in the electron shells. These vacancies will be filled by electrons from outer shells resulting in either emission of a fluorescence x-ray, also called characteristic x-ray, or an Auger electron. The probability that the vacancy results in an x-ray is called the fluorescence yield. The energy of the fluorescent x-ray is given by the difference in the binding energy of the shell the electron comes from and the binding energy of the shell where the vacancy is. Previously fluorescence x rays had different notations according to a system developed by Siegbahn, a Swedish physicist. Lately it has become more common to just indicate the shells. E.g., if the vacancy is in the K-shell and the electron comes from the L_3-shell, the x-ray was called a $K_{\alpha 1}$-x-ray in the Siegbahn notation. However, it is more clear if it is just called x_{K-L_3}. The energies and the fluorescence yields of different fluorescence x rays can e.g. be found in the Table of Radioactive Isotopes (Browne and Firestone, 1986).

As the de-excitation results in either a γ-ray or an internal conversion electron, the following equations are obtained

$$\alpha = \frac{N_e}{N_\gamma} \tag{1.1.16}$$

$$N_e + N_\gamma = 1 \tag{1.1.17}$$

This gives

$$N_\gamma = \frac{1}{1 + \alpha} \tag{1.1.18}$$

The number of internal conversion electrons from the different shells are obtained through the relation $N_{e_{K,L,M...}} = \alpha_{K,L,M...} \cdot N_\gamma$.

From Tables of Isotopes (Lederer and Shirley, 1978) the following data are obtained for ^{137m}Ba. (There are several different sets of data for the internal conversion coefficients in the table and these are just one example).

$\alpha_K = 0.0916$, $\alpha_L = 1.162 \cdot 10^{-2}$, $\alpha_M = 4.2 \cdot 10^{-3}$, $\alpha_N = 0$

$\alpha = \alpha_K + \alpha_L + \alpha_M$

giving α=0.1124

This gives

$$N_\gamma = \frac{1}{1 + 0.1124} = 0.899$$

and

$N_{e_K} = 0.0916 \cdot 0.899 = 0.0823$, $N_{e_L} = 0.0162 \cdot 0.899 = 0.0146$, $N_{e_M} = 4.2 \cdot 10^{-3} \cdot 0.899 = 3.8 \cdot 10^{-3}$.

The energies of the internal conversion electrons are given by

$E_{e_K} = E_\gamma - B_K$=0.6616 – 0.0374 = 0.6242 MeV
$E_{e_L} = E_\gamma - B_K$=0.6616 – 0.0052 = 0.6560 MeV (L_2-shell)
$E_{e_M} = E_\gamma - B_K$=0.6616 – 0.0011 = 0.6605 MeV (M_2-shell).

For e_L and e_M only one energy is included as the difference in energy between the subshells is small. The data are summarized in Table 1.2.

The number of x rays and their energies are obtained by e.g. *Table 7* in Table of Radioactive Isotopes (Eckerman and Endo, 1986). Table 1.3 shows data taken from that table giving the number of x_K-rays per 100 vacancies and their energies. Only the most frequent x rays are included in the table. The last column shows the number of x rays

Table 1.2: Frequency and energy of the photons and internal conversion electrons from the ^{137}Cs-decay.

Particle	Frequency	Energy (MeV)
γ	0.899	0.6616
e_K	0.0823	0.624
e_L	0.015	0.656
e_M	0.0038	0.660

from the de-excitation of ^{137m}Ba, where column two has been multiplied with 0.0823, the number of K-shell vacancies. As shown the probabilities are larger for K-L transitions except for the transition K-L_1 which is called forbidden.

Table 1.3: Frequency and energy of the fluorescence x rays emitted in the ^{137}Cs-decay.

x-ray	Frequency	Energy (keV)	x rays per de-excitation
$K - L_3$	0.467	32.194	0.0384
$K - L_2$	0.256	31.817	0.0211
$K - L_1$	0.0000334	31.452	0
$K - M_3$	0.0863	36.378	0.0071
$K - N_2N_3$	0.0273	37.255	0.0022
$K - M_2$	0.0447	36.304	0.0037

There are also x rays from vacancies in the L-shell. These vacancies are due to both the internal conversion processes in the L-shell and to vacancies produced when vacancies in the K-shell are filled with electrons from the L-shell, resulting in new vacancies in the L-shell. Table of Radioactive Isotopes, *Table 7* includes data for both these possibilities. However, due to the sub-shells in the outer electron shells, there will now be a fine structure of x-ray photons with different energies. The energy of the L x rays is normally rather low, in this case around 5 keV, and often it is possible to average all energies with one energy in particular as the frequency also is low. In this de-excitation there will be 0.0067 (0.0823·0.081) L x rays due to the vacancies in the K-shell and 0.0014 (0.0146·0.095) L x rays from the L-vacancies from the internal conversion per de-excitation. The energies are between 4 and 6 keV. Note that the decay started with a ^{137}Cs nuclide, but the data for the x rays must be calculated for ^{137m}Ba, as it is from the Ba-nuclide the de-excitation is due.

An alternative to fluorescence x rays are Auger electrons. These are produced when an electron from an outer shell goes to an inner shell and the excess energy obtained is used to emit another electron. If the process is due to energy transfers between subshells the emitted electrons are often called Coster-Cronig electrons. When there is a vacancy in the K-shell, the most common Auger process is when an electron goes from the L-shell and there is an Auger-electron emitted also from the L-shell (A_{KLL}-electron).

The energy of the Auger electrons are obtained as the energy difference in the binding energies between the electron shells. It is important to include the binding energy of the emitted Auger electron. E.g. in the process with an A_{KLL}-electron, the electron energy will be $E_{KLL}=B_K - B_L - B_L$.

The Auger process is an alternative to the emission of fluorescence x rays. The probability for the two different processes varies with the atomic number. For low atomic numbers the Auger electrons are dominating and for high atomic numbers the fluorescence x rays. Data for the probabilities are found in the Table of Radioactive Isotopes, *Table 8*. In Table 1.4 data for the five most common Auger electrons from ^{137m}Ba de-excitation are shown. In the original *Table 8* there are probabilities for 38 different combinations of shell transitions through vacancies in the K-shells giving Auger electrons.

Table 1.4: Frequency and energy of the Auger electrons emitted in the ^{137}Cs-decay.

Auger electron	Frequency	Energy (keV)	Auger electrons per de-excitation
	$(\times 10^{-2})$		$(\times 10^{-2})$
$K - L_1 L_1$	0.79	25.46	0.065
$K - L_1 L_2$	0.97	25.83	0.080
$K - L_1 L_3$	1.11	26.21	0.091
$K - L_2 L_3$	2.35	26.57	0.19
$K - L_3 L_3$	1.17	26.95	0.096

The Auger electrons and also the fluorescence x rays will produce new vacancies in the outer shells and will give rise to a cascade of mainly Auger- and Coster-Cronig-electrons with low energies.

In the calculations we have separated the decay from ^{137}Cs and the de-excitation from ^{137m}Ba. As mentioned above normally there is radiation secular equilibrium, and when the decay from ^{137}Cs is described the de-excitation from ^{137m}Ba is included. Then all figures above for ^{137m}Ba have to be multiplied with the probability that ^{137}Cs goes to the excited level ^{137m}Ba, that is 0.946. Thus e.g. the number of 0.6616 MeV γ-rays per ^{137}Cs decay is 0.946·0.899=0.85. When using published data of the ^{137}Cs decay it is important to check how this is handled as this could be different in different tables.

1.2 Exercises in Radiation Sources and Radioactive Decay

Exercise 1.1. The α-decay of a nucleus with a mass number of 200 has two components with energies 4.687 MeV and 4.650 MeV. Neither of the decays goes to the basic level of the daughter, but is accompanied with emission of γ-radiation, with the respective energies 266 and 305 keV. No other γ-energies are found. Construct from this information the decay scheme.

Exercise 1.2. A radioactive source (^{210}At) with unknown activity is positioned in a vacuum chamber with the volume 10.0 dm^3. The activity is determined by measuring the amount of He that is obtained when the emitted α-particles attract electrons and produce stable He-atoms. Assume that all α-particles are transformed to He-gas. After 24.0 d the He-mass is determined to 1.84 ng. Calculate the initial activity of ^{210}At.

Exercise 1.3. ^{239}Pu is α-radioactive and the α-particles are emitted with a kinetic energy of 5.144 MeV. When a sphere of ^{239}Pu, with a mass of 120.06 g, is placed in a calorimeter with liquid nitrogen, the same mass of nitrogen is vaporized per time unit as when 0.231 W is added by electric means. Calculate the half life of ^{239}Pu.

Exercise 1.4. A radioactive source of pure ^{210}Bi, is at time t=0, placed in a container of lead, that absorbs all emitted radiation including the produced bremsstrahlung, when the electrons are absorbed. All energy, that is absorbed in the radioactive source and the container is transferred to heat. The heat power is measured with a calorimeter.

Calculate the time it takes until the heat power has decreased to 3.60 mW. At time t=0 the ^{210}Bi activity is 90.0 GBq. Assume that all decays go to the ground state of the daughter nuclide and that only one particle is emitted in every decay.

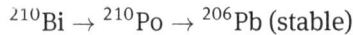

^{210}Bi $\rightarrow$ ^{210}Po $\rightarrow$ ^{206}Pb (stable)
^{210}Bi : $T_{1/2}$ =5.01 d, $E_{\beta max}$ =1.16 MeV, $E_{\beta mean}$ =0.344 MeV
^{210}Po : $T_{1/2}$ =138.4 d, E_{α} =5.2497 MeV

Exercise 1.5. A ^{137}Cs-source is measured and a count rate of $0.45 \cdot 10^6$ pulses per minute is obtained for photons with the energy 0.662 MeV. How many counts per minute can be expected for photons with an energy between 30 and 40 keV? Assume the detector measures every 10th photon independent of energy. Calculate also the activity of the source.

Exercise 1.6. The γ-photons and the fluorescence x rays obtained when ^{125}I disintegrates, may be used for x-ray imaging. How many photons are obtained per decay?

$\frac{EC_L}{EC_K}$ =0.23, $EC_M = 0$

α=13.8, α_K=11.7, α_M=0

where EC_L/EC_K is the ratio between the probabilities that a nucleus captures an L- and a K-electron and α is the internal conversion coefficient.

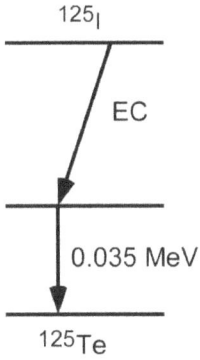

Figure 1.2: Decay scheme for ^{125}I.

Exercise 1.7. ^{111}In is a radionuclide which is used for investigations in nuclear medicine. It is then possible to use both the emitted γ-radiation and the fluorescence x rays. Calculate the ratio between the count rates obtained with one energy window at 245±10 keV and one at 25±5 keV. Assume that the detector efficiency for 245 keV is 12% and for 25 kev 15%. 85% of the electron capture belongs to the K-shell.

γ_1 : α_K= 0.100, α_K/α_L = 6.22, α_M = 0
γ_2 : α_K= 0.0540, α_K/α_L = 5.20, α_M = 0
$\alpha_{K,L,M}$= internal conversion coefficients for K, L and M shells respectively
$\alpha_K = N_{e_K}/N_\gamma$
N_{e_K}=number of emitted internal conversion electrons from the K-shell per decay
N_γ=number of emitted γ-photons per decay.

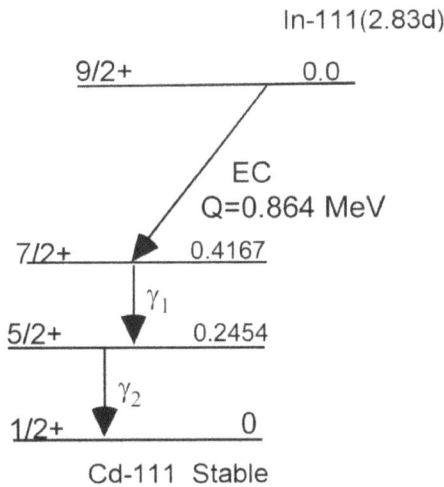

Figure 1.3: Decay scheme for ^{111}In.

Exercise 1.8. ^{241}Pu can be produced at nuclear reprocessing plants and con-taminate the environment. A sample from the biosphere is measured at a certain date and the ratio of α-activity of ^{241}Am/^{239}Pu is determined to $2.00 \cdot 10^{-3}$. The sample was collected 5.0 years ago. The corresponding ratio was then $0.90 \cdot 10^{-3}$. Calculate the α-activity ratio ^{241}Pu/^{239}Pu at time of collection.

Exercise 1.9. ^{104}Rh can be obtained through irradiation of ^{103}Rh with thermal neutrons. The following nuclear reactions are obtained

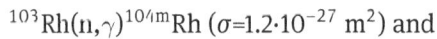

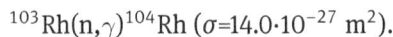

^{103}Rh$(n,\gamma)^{104m}$Rh $(\sigma = 1.2 \cdot 10^{-27}$ m$^2)$ and

^{103}Rh$(n,\gamma)^{104}$Rh $(\sigma = 14.0 \cdot 10^{-27}$ m$^2)$.

^{104m}Rh decays through internal transition to ^{104}Rh. The half lives are ^{104m}Rh: $T_{1/2} = 4.40$ min and ^{104}Rh: $T_{1/2} = 42$ s.

Calculate the activity of ^{104}Rh in percentage of the saturation activity 60 s and 600 s after start of irradiation.

Exercise 1.10. The radionuclide ^{169}Yb is obtained through neutron irradiation of ^{168}Yb. The interaction cross section is large, $1.10 \cdot 10^{-24}$ m^2. The half-life of the produced radionuclide is 32.0 d. Calculate the time when the activity is maximal, if the neutron fluence rate is $1.00 \cdot 10^{18}$ m^{-2}s^{-1}.

Exercise 1.11. A ^{99}Mo-^{99m}Tc generator has the activity 12.0 GBq Monday morn-ing at 8:00. Thursday morning at 8:00 the generator is completely emptied of ^{99m}Tc. Later the same day the generator is eluted again and completely emptied of ^{99m}Tc. By mistake the time for the elution was not observed. The activity of the eluted ^{99m}Tc is measured later and is 2.10 GBq at 16:00 the same day. When was the generator eluted?

Exercise 1.12. When running a reactor the fission product ^{149}Nd is obtained. This disintegrates to ^{149}Sm according to the reactions below.

^{149}Nd$(\beta, T_{1/2} = 1.70$ h$) \rightarrow ^{149}$Pm$(\beta, T_{1/2} = 47.0$ h$) \rightarrow ^{149}$Sm

^{149}Sm is stable but disappears during the drift of the reactor through the reac-tion

^{149}Sm$+$n$\rightarrow ^{150}$Sm

When the reactor is shut down, this reaction will not continue and the amount of ^{149}Sm will increase. This may be a problem because the neutron capture cross-section of ^{149}Sm is much larger than the fission cross-section for ^{235}U and it may be difficult to restart the reactor if there is a lot of samarium in the fuel.

If there is 300 TBq ^{149}Nd and 1000 TBq ^{149}Pm when the reactor is shut down, what mass of ^{149}Sm will be produced during the first 48 h after the shut down? (The activity numbers are just examples and not meant to be realistic).

1.3 Solutions in Radiation Sources and Radioactive Decay

Solution exercise 1.1.
In an α-decay the transition energy, Q, is divided between the α-particle and the daughter nuclide. If the daughter nuclide has an excited state with the energy E_γ, not all transition energy will be kinetic energy of the two particles but instead, $Q' = Q - E_\gamma$.

The relation between Q' and the kinetic energy of the α-particle, E_α, is given by

$$Q' = \frac{m_\alpha + m_D}{m_D} E_\alpha \qquad (1.3.1)$$

Data:
$m_\alpha = 4.00$ (mass of α-particle)
$m_D = 196$ (mass of daughter nuclide)
$E_{\alpha 1} = 4.687$ MeV (kinetic energy of α-particle 1)
$E_{\alpha 2} = 4.650$ MeV (kinetic energy of α-particle 2)
$E_{\gamma 1} = 0.266$ MeV (energy of γ ray 1)
$E_{\gamma 2} = 0.305$ MeV (energy of γ ray 2)

Inserting the masses and energies in Eq. 1.3.1 gives

$Q'_1 = 4.783$ MeV, $Q'_2 = 4.745$ MeV

From this information it is now possible to construct the decay-scheme (Fig. 1.4).

Solution exercise 1.2.
^{210}At is decaying to ^{210}Po by electron capture. ^{210}Po is also radioactive and decays by α-decay to ^{206}Pb, which is stable. The activity of ^{210}Po is given by the equation

$$A_{Po}(t) = \frac{A_{At}(0)\lambda_{Po}}{\lambda_{Po} - \lambda_{At}} \left(e^{-\lambda_{At}t} - e^{-\lambda_{Po}t} \right) \qquad (1.3.2)$$

The number of ^{210}Po disintegrations during a time T is obtained by integrating equation (1.3.2).

$$N = \int_0^T A_{Po}(t)\, dt = \frac{A_{At}(0)\lambda_{Po}}{\lambda_{Po} - \lambda_{At}} \left[-\frac{e^{-\lambda_{At}t}}{\lambda_{At}} + \frac{e^{-\lambda_{Po}t}}{\lambda_{Po}} \right]_0^T \qquad (1.3.3)$$

$$N = \frac{A_{At}(0)\lambda_{Po}}{\lambda_{Po} - \lambda_{At}} \left[\frac{e^{-\lambda_{Po}T}}{\lambda_{Po}} - \frac{e^{-\lambda_{At}T}}{\lambda_{At}} - \frac{1}{\lambda_{Po}} + \frac{1}{\lambda_{At}} \right] \qquad (1.3.4)$$

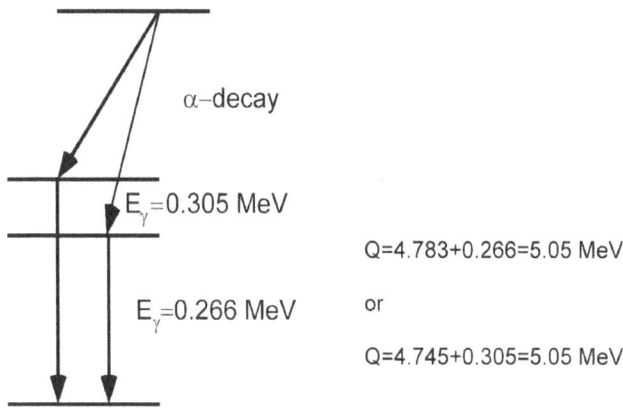

Figure 1.4: Proposed decay scheme.

This is equal to the number of α-particles, as there is one α-particle per decay. Every α-particle is capturing electrons and is transferred to a He-nucleus. The relation between the number of He-nuclei N_{He} and mass is given by the equation

$$N_{\text{He}} = \frac{m_{\text{He}} N_A}{m_a} \tag{1.3.5}$$

where
$N_A = 6.022 \cdot 10^{26}$ atoms/kmol (Avogadro's number)
$m_{\text{He}} = 1.84 \cdot 10^{-12}$ kg (He-mass)
$m_a = 4.00$ (atomic mass of He)

Decay data:
$\lambda_{\text{At}} = \ln 2/8.10 = 0.08557\ \text{h}^{-1} = 2.0538\ \text{d}^{-1}$
$\lambda_{\text{Po}} = \ln 2/138.38 = 5.009 \cdot 10^{-3}\ \text{d}^{-1}$
$T = 24\ \text{d}$

Data inserted gives

$$N_{\text{He}} = \frac{1.84 \cdot 10^{-12} \cdot 6.022 \cdot 10^{26}}{4.00} = 2.770 \cdot 10^{14}$$

and

$$2.770 \cdot 10^{14} = \frac{A_{\text{At}}(0) \cdot 5.009 \cdot 10^{-3} \cdot 24 \cdot 3600}{5.009 \cdot 10^{-3} - 2.0538}$$
$$\times \left[\frac{e^{-5.009 \cdot 10^{-3} \cdot 24}}{5.009 \cdot 10^{-3}} - \frac{e^{-2.0538 \cdot 24}}{2.0538} - \frac{1}{5.009 \cdot 10^{-3}} + \frac{1}{2.0538} \right]$$

$A_{At}(0)$=59.2 GBq

Answer: The initial activity of ^{210}At is 59 GBq.

Solution exercise 1.3.

In a radioactive decay the energy released per time unit, P, is given by the relation

$$P = \lambda N Q \tag{1.3.6}$$

where λ is the decay constant, N is $(N_A m)/(m_a)$ (number of nuclides) and Q is the disintegration energy.

This gives an expression for the half life $T_{1/2}$ $(=\ln 2/\lambda)$

$$T_{1/2} = \frac{\ln 2 \, N_A \, m \, Q}{P \, m_a} \tag{1.3.7}$$

Data:

m =120.06 g

$N_A = 6.022 \cdot 10^{26}$ atoms/kmol (Avogadro's number)

m_{aPu}=239.05 $(^{239}$Pu$)$

P=0.231 W (power by electric means, that should be the same as by radioactive decay)

When a radionuclide disintegrates through α-decay, the disintegration energy is divided between the α-particle and the daughter nuclide (see Fig. 1.5). From the kinetic energy and the momentum relations Q can be obtained:

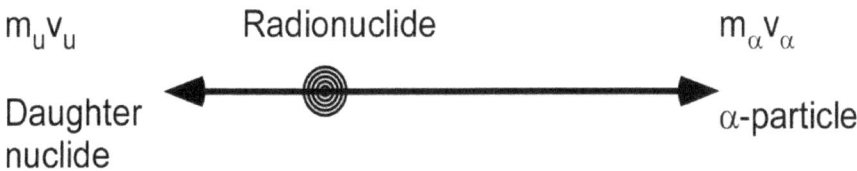

$m_u v_u$ Radionuclide $m_\alpha v_\alpha$

Daughter nuclide α-particle

Figure 1.5: Mechanics of the α-decay. Conservation of momentum must hold in the decay process.

Kinetic energy:

$$Q = \frac{m_D \, v_D^2}{2} + \frac{m_\alpha \, v_\alpha^2}{2} \tag{1.3.8}$$

Momentum:

$$m_D \, v_D = m_\alpha \, v_\alpha \tag{1.3.9}$$

Equations (1.3.8) and (1.3.9) give

$$Q = \frac{m_D \, v_\alpha^2 \, m_\alpha^2}{2 m_D^2} + \frac{m_\alpha \, v_\alpha^2}{2} = \frac{m_\alpha}{m_D} E_\alpha + E_\alpha \tag{1.3.10}$$

Observe that it is the atomic mass of the daughter nuclide that should be inserted in Eq. 1.3.10, and thus $m_\alpha = 4.003$ and $m_D = 235.04$. The kinetic energy of the α-particle is given in the problem, $E_\alpha = 5.144$ MeV. Data inserted in Eq. 1.3.10 gives

$$Q = \frac{4.003}{235.04} \cdot 5.144 + 5.144 = 5.2316 \, \text{MeV}$$

Data inserted in Eq. (1.3.7) gives

$$T_{1/2} = \frac{\ln 2 \cdot 6.022 \cdot 10^{26} \cdot 5.2316 \cdot 1.602 \cdot 10^{-13} \cdot 0.12006}{0.231 \cdot 239.05}$$

$T_{1/2} = 7.606 \cdot 10^{11}$ s $= 24.1 \cdot 10^3$ years.

Answer: The half life of ^{239}Pu is $24.1 \cdot 10^3$ years

Solution exercise 1.4.
The activity of ^{210}Bi, $A_{Bi}(t)$, varies with time according to the relation

$$A_{Bi}(t) = A_{Bi}(0)e^{-\lambda_{Bi}t} \tag{1.3.11}$$

The activity of ^{210}Po, $A_{Po}(t)$, varies with time according to the relation

$$A_{Po}(t) = \frac{\lambda_{Po}A_{Bi}(0)}{\lambda_{Po} - \lambda_{Bi}}(e^{-\lambda_{Bi}t} - e^{-\lambda_{Po}t}) \tag{1.3.12}$$

The power from the radioactive nuclides is obtained from the product of the activity and the energy released per decay. At time t there is activity both from ^{210}Bi and ^{210}Po present. Thus the total power at time t is given by

$$P = A_{Bi}(0)\bar{E}_\beta e^{-\lambda_{Bi}t} + \frac{\lambda_{Po}A_{Bi}(0)Q}{\lambda_{Po} - \lambda_{Bi}}(e^{-\lambda_{Bi}t} - e^{-\lambda_{Po}t}) \tag{1.3.13}$$

Q is the total energy released in the α decay. The kinetic energy both from the α-particle and the daughter nucleus must be included in the calculations. The total energy is obtained from the equation

$$Q = E_D + E_\alpha = \frac{E_\alpha m_\alpha}{m_D} + E_\alpha \tag{1.3.14}$$

Data:
$A_{Bi}(0) = 90.0$ GBq
$m_\alpha = 4.00$ (atomic mass of He)
$m_D = 206$ (atomic mass of ^{206}Pb)
$E_\alpha = 5.2497$ MeV $\Rightarrow Q = 5.3516$ MeV (disintegration energy)
$\bar{E}_\beta = 0.344$ MeV (mean energy of β-radiation from ^{210}Bi)
$P = 3.60$ mW (electric power)

λ_{Bi}=ln2/5.01=0.1384 d^{-1} (decay constant for ^{210}Bi)
λ_{Po}=ln2/138.4=5.009·10^{-3} d^{-1} (decay constant for ^{210}Po)

Data inserted in Eq. (1.3.13) gives

$$3.60 \cdot 10^{-3} = 90.0 \cdot 10^{9} \cdot 0.344 \cdot 1.602 \cdot 10^{-13} e^{-0.1384t} +$$
$$\frac{5.009 \cdot 10^{-3} \cdot 90.0 \cdot 10^{9} \cdot 5.3516 \cdot 1.602 \cdot 10^{-13}}{5.009 \cdot 10^{-3} - 0.1384} (e^{-0.1384t} - e^{-5.009 \cdot 10^{-3}t})$$

This is simplified to

$$3.60 \cdot 10^{-3} = 2.06 \cdot 10^{-3} e^{-0.1384t} + 2.90 \cdot 10^{-3} e^{-5.009 \cdot 10^{-3}t}$$

This equation can be solved numerically giving, t=6.84 d.

Answer: The time for the heat power to decrease to 3.60 mW is 6.84 d.

Solution exercise 1.5.
^{137}Cs disintegrates to ^{137m}Ba. The de-excitation from the metastable level ^{137m}Ba results in the measured photons. The result is given in Table 1.5.

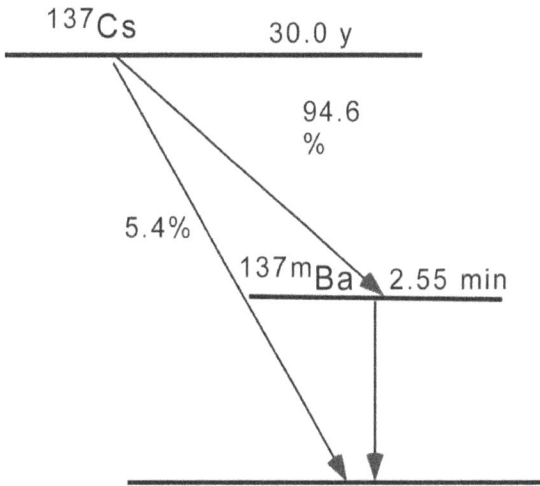

Figure 1.6: Decay scheme for ^{137}Cs.

The count rate r_{30-40} in channels 30-40 keV is given by

$r_{30-40} = r_{\gamma} \cdot f_{30-40}/f_{\gamma}$

Assume the same efficiency for the two photon energies. Data inserted gives

Table 1.5: De-excitation data for $^{137\text{m}}$Ba and measured count rate. Data for the decay are taken from Table of Radionuclide Transformations (ICRP, 1986).

Photons	Frequency(photons per de-excitation)	Count rate
Gamma-rays	f_γ=0.898	r_γ=0.45·10^6 min^{-1}
Fluorescence x rays	f_{30-40}=0.0605 (K$_\alpha$+K$_\beta$)	??

$$r_{30-40} = \frac{0.45 \cdot 10^6 \cdot 0.0605}{0.898} = 30317 \, \text{min}^{-1}$$

To calculate the activity of ^{137}Cs one has to know how many excited $^{137\text{m}}$Ba-nuclei there are for each ^{137}Cs decay. Table of Radionuclide Transformations (ICRP, 1986) gives that the relative number of decays going to the excited level $^{137\text{m}}$Ba is 0.946. This gives f'_γ=0.946·0.898 γ-rays per ^{137}Cs decay.

The count rate is given by the relation

$$r_\gamma = A f'_\gamma n_\gamma \qquad (1.3.15)$$

where A is the activity of the ^{137}Cs source and η_γ=0.10 is the efficiency of the detector.
This gives the activity

$$A = \frac{r_\gamma}{f'_\gamma \eta_\gamma} = \frac{0.45 \cdot 10^6}{60 \cdot 0.10 \cdot 0.946 \cdot 0.898}$$

$A = 88.3 \cdot 10^3$ Bq.

Answer: The count rate for photons between 30 and 40 keV is 30.3·10^3 min^{-1}. The ^{137}Cs-activity is 88 kBq.

Solution exercise 1.6.

^{125}I disintegrates by electron capture to an excited state of ^{125}Te. The electron capture will give rise to an electron vacancy in the K- or L-electron shells followed by emission of fluorescence x rays and/or Auger electrons. The de-excitation from the excited state of ^{125}Te results in emission of γ-radiation and/or internal conversion electrons. The vacancies obtained after the emission of internal electrons will also produce emission of fluorescence x rays and/or Auger electrons.

The decay data are given in Fig. 1.7 and EC_L/EC_K is the ratio between the probabilities that a nucleus captures an L- or a K-electron and α is the total internal conversion coefficient.

125I

EC   ECK/ECL=0.23, ECM=0

0.035 MeV α_T=13.8, α_K=11.7

125Te

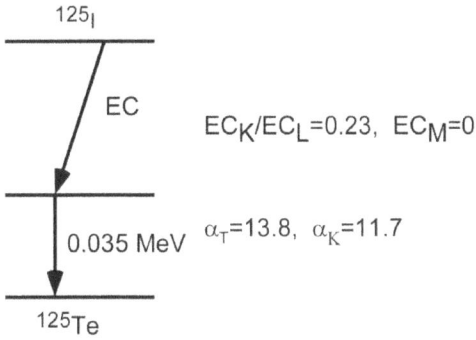

Figure 1.7: Decay scheme for 125I.

Electron capture:
Combining the relations (as $EC_M=0$, the electron capture is either from the K- or the L-shell)

$$\begin{cases} \frac{EC_L}{EC_K} = 0.23 \\ EC_K + EC_L = 1 \end{cases}$$

results in

EC_K=0.813 and EC_L=0.187.

This means that on average there are 0.813 electron captures from the K-shell and 0.187 from the L-shell.

De-excitation:
Combining the relations (there is either a gamma ray or an internal conversion electron) gives

$$\begin{cases} \alpha = \frac{N_e}{N_\gamma} \\ N_e + N_\gamma = 1 \end{cases}$$

and

$$N_\gamma = \frac{1}{\alpha + 1}$$

Data inserted gives N_γ =0.0676.

The number of internal conversion electrons from the K-shell N_{e_K} is then given by

$$N_{e_K} = \alpha_K \, N_\gamma = 11.7 \cdot 0.0676 = 0.791$$

Total:
The total number of K-electrons, N_{e_K}, and thus vacancies in the K-shell, from both the electron capture and the de-excitation is then given by

$$N_{e_{K,tot}} = 0.813 + 0.791 = 1.604$$

The number of fluorescence K x rays is given by the relation

$$N_{X_K} = N_{e_{K,tot}} \omega_K$$

$\omega_K = 0.877$ (fluorescence yield)

Observe that ω_K shall be taken for tellurium as this is the atom after the decay.

This gives

$$N_{X_K} = 0.877 \cdot 1.604 = 1.407$$

To obtain the total number of photons per decay, the number of gamma rays, N_γ, and the number of fluorescence K x rays, N_{X_K}, are added.

$$N = 1.407 + 0.0676 = 1.4746$$

Answer: The total number of photons per 100 decays is 147.

Solution exercise 1.7.
The photons in the energy interval 25±5 keV are fluorescence K-x rays from cadmium. The photons in the energy interval 245±10 keV come from the de-excitation from the 0.2454 MeV level. The de-excitation can be performed either by emitting a gamma photon or an internal conversion electron. The electron will leave a vacancy in a electron shell, which will be filled by electrons from outer shells and give rise to emission of fluorescence x rays and/or Auger electrons.

The relation between gamma photons and internal conversion electrons is given by the internal conversion coefficient $\alpha = N_e / N_\gamma$.

In-111(2.83d)

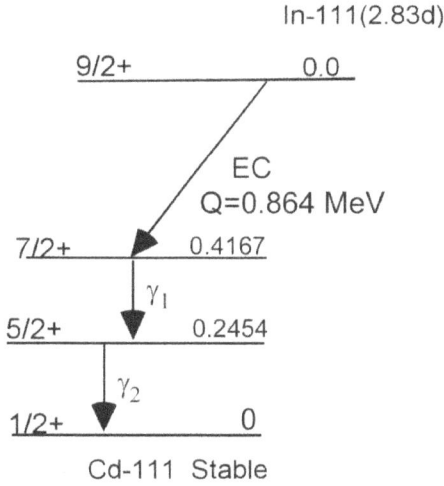

Figure 1.8: Decay scheme for ^{111}In.

Energy interval 245±10 keV

The number of N_γ photons per decay is obtained from the equations

$$
\begin{cases}
\alpha = \alpha_K + \alpha_L = \frac{N_e}{N_\gamma} \\
N_e + N_\gamma = 1
\end{cases}
$$

which gives

$$N_\gamma = \frac{1}{1+\alpha}$$

Energy interval 25±5 keV

The number of electron vacancies in the K-shell is obtained from internal conversion both for γ_1 and γ_2. The number of internal conversion electrons from the K-shell is obtained from the relation $N_{eK_\gamma} = \alpha_K N_\gamma$. There are also electron vacancies from the electron capture, N_{eKEC}. This number is given in the exercise text.

The total number of electron vacancies per decay, N_{eK}, is now obtained by adding the different components.

$$N_{eK} = N_{eK_{\gamma_1}} + N_{eK_{\gamma_2}} + N_{eKEC}$$

The number of K-fluorescence x rays is obtained by multiplying with the fluorescence yield ω_K.

$$N_{X_K} = \omega_K N_{eK}$$

Data:

$\gamma_1 : \alpha_{K1} = 0.100,\ \alpha_{K1}/\alpha_{L1} = 6.22,\ \alpha_{M1} = 0$
$\gamma_2 : \alpha_{K2} = 0.0540,\ \alpha_{K2}/\alpha_{K2} = 5.20,\ \alpha_{M2} = 0$
$N_{e_{KEC}} = 0.85$
$\omega_K(Cd) = 0.843$

Data inserted in the relations above gives

$\alpha_1 = 0.100 + 0.100/6.22 = 0.1161$ and $N_{\gamma1} = 0.896$

This will give the number of electron vacancies in the K-shell.

$N_{e_{K1}} = 0.896 \cdot 0.100 = 0.0896$

Correspondingly for γ_2 is obtained

$\alpha_2 = 0.054 + 0.054/5.20 = 0.06438$ and $N_{\gamma2} = 0.9395$

$N_{e_{K2}} = 0.9395 \cdot 0.054 = 0.0507$

Total number of K-vacancies is thus

$N_{e_K} = 0.0896 + 0.0507 + 0.85 = 0.9903$

Total number of K x rays $N_{X_K} = 0.9903 \cdot 0.843 = 0.8348$

The count ratio between the number of photons of energy 245 keV and energy 25 keV is given by the relation

$$\frac{N_{\gamma2}\eta_{245}}{N_{X_K}\eta_{25}} \tag{1.3.16}$$

where $\eta_{245} = 0.12$ and $\eta_{25} = 0.15$ are the detector efficiencies for the different energies.

This gives the ratio in count rate

$$\frac{0.9395 \cdot 0.12}{0.8348 \cdot 0.15} = 0.900$$

Answer: The count rate ratio is 0.90.

Solution exercise 1.8.
^{241}Pu decays to ^{241}Am by β-decay. ^{241}Am is also radioactive and decays by α-decay.

^{241}Pu(β,T$_{1/2}$=14.4 y)$\rightarrow$^{241}Am(α,T$_{1/2}$=433 y)$\rightarrow$^{237}Np$\rightarrow$...

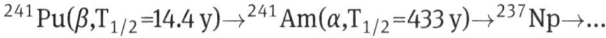

^{239}Pu is radioactive with a half-life T$_{1/2}$=2.41·10^4 y. This means that the decay of ^{239}Pu can be neglected in the calculations as the time period considered is five years.
The activity of ^{241}Am produced from ^{241}Pu after t years is given by the relation

$$A_{\text{Am},1}(t) = \frac{\lambda_{\text{Am}}A_{\text{Pu}}(0)}{\lambda_{\text{Am}} - \lambda_{\text{Pu}}}(e^{-\lambda_{\text{Pu}}t} - e^{-\lambda_{\text{Am}}t}) \qquad (1.3.17)$$

where
$A_{\text{Pu}}(0)$= activity of ^{241}Pu at t=0
λ_{Pu}=ln2/14.4=4.8135·10^{-2} y^{-1} (decay constant for ^{241}Pu)
λ_{Am}=ln2/433=1.601·10^{-3} y^{-1} (decay constant for ^{241}Am)

According to information in the problem, the activity of ^{241}Am at t=0 is given by the relation

$A_{\text{Am}}(0) = 0.90 \cdot 10^{-3}A_3$

A_3=activity of ^{239}Pu (assumed to be constant).

After t years the activity of ^{241}Am, A_{Am}, has decreased due to radioactive decay.

$A_{\text{Am},2}(t) = 0.90 \cdot 10^{-3}A_3 \cdot e^{-\lambda_{\text{Am}}t}$

Thus the total activity of ^{241}Am after t years is

$$A_{\text{Am}}(t) = \frac{\lambda_{\text{Am}}A_{\text{Pu}}(0)}{\lambda_{\text{Am}} - \lambda_{\text{Pu}}}(e^{-\lambda_{\text{Pu}}t} - e^{-\lambda_{\text{Am}}t}) + 0.90 \cdot 10^{-3}A_3 \cdot e^{-\lambda_{\text{Am}}t} \qquad (1.3.18)$$

If t= 5.0 y then $A_{\text{Am}}(5) = 2.0 \cdot 10^{-3}A_3$ according to measurements.

Inserting λ_{Pu}, λ_{Am} and t=5.0 y in Eq. (1.3.18) gives

$$2.0 \cdot 10^{-3}A_3 = \frac{1.601 \cdot 10^{-3}A_{\text{Pu}}(0)}{1.601 \cdot 10^{-3} - 4.8135 \cdot 10^{-2}}(e^{-5 \cdot 4.8135 \cdot 10^{-2}} - e^{-5 \cdot 1.601 \cdot 10^{-3}})$$
$$+ 0.90 \cdot 10^{-3}A_3 \cdot e^{-5 \cdot 1.601 \cdot 10^{-3}}$$

Solving Eq. (1.3.18) gives

$A_{\text{Pu}}(0) = 0.156A_3$

Answer: The α-activity ratio A(^{241}Pu)/A(^{239}Pu) is 0.16 at time of collection of the sample.

Solution exercise 1.9.

Production of ^{104m}Rh-nuclei (N_1) is given by Eq. (1.3.19), if it is assumed that the decrease in the number of the ^{103}Rh-nuclei may be neglected.

$$\frac{dN_1}{dt} = U_1 - \lambda_1 N_1 \tag{1.3.19}$$

where dN_1/dt is the change in the number of N_1-nuclei per time unit, U_1 is the production rate in the reaction giving ^{104m}Rh and λ_1 is the decay constant of nuclide N_1.

The solution of Eq. (1.3.19) is

$$N_1 = \frac{U_1}{\lambda_1}(1 - e^{-\lambda_1 t}) \tag{1.3.20}$$

The production of ^{104}Rh-nuclei (N_2) through activation and decay from the excited state of ^{104m}Rh is given by the equation

$$\frac{dN_2}{dt} = U_2 + \lambda_1 N_1 - \lambda_2 N_2 \tag{1.3.21}$$

Eq. (1.3.20) combined with Eq. (1.3.21) gives

$$\frac{dN_2}{dt} = U_2 - \lambda_2 N_2 + U_1(1 - e^{-\lambda_1 t}) \tag{1.3.22}$$

Multiply with integrating factor $e^{\lambda_2 t}$

$$\frac{dN_2}{dt}e^{\lambda_2 t} = U_2 e^{\lambda_2 t} - \lambda_2 N_2 e^{\lambda_2 t} + U_1(1 - e^{-\lambda_1 t})e^{\lambda_2 t} \tag{1.3.23}$$

Eq. (1.3.23) can be rewritten as

$$\frac{d(N_2 e^{\lambda_2 t})}{dt} = U_2 e^{\lambda_2 t} + U_1(1 - e^{-\lambda_1 t})e^{\lambda_2 t} \tag{1.3.24}$$

Eq. (1.3.24) is integrated

$$\int \frac{d(N_2 e^{\lambda_2 t})}{dt}\, dt = \int \left[U_2 + U_1(1 - e^{-\lambda_1 t}) \right] e^{\lambda_2 t}\, dt \tag{1.3.25}$$

The indefinite solution of the integral gives

$$N_2 e^{\lambda_2 t} = \frac{(U_2 + U_1)e^{\lambda_2 t}}{\lambda_2} - \frac{U_1 e^{(\lambda_2 - \lambda_1)t}}{\lambda_2 - \lambda_1} + C \tag{1.3.26}$$

C can be determined by assuming that at $t=0$, $N_2(0)=0$. This inserted in Eq. (1.3.26) gives

$$0 = \frac{(U_2 + U_1)}{\lambda_2} - \frac{U_1}{\lambda_2 - \lambda_1} + C \tag{1.3.27}$$

$$C = -\frac{(U_2 + U_1)}{\lambda_2} + \frac{U_1}{\lambda_2 - \lambda_1} \tag{1.3.28}$$

Combining this with Eq. (1.3.26) results in

$$N_2 e^{\lambda_2 t} = \frac{(U_2 + U_1)e^{\lambda_2 t}}{\lambda_2} - \frac{U_1 e^{(\lambda_2 - \lambda_1)t}}{\lambda_2 - \lambda_1} - \frac{(U_2 + U_1)}{\lambda_2} + \frac{U_1}{\lambda_2 - \lambda_1} \tag{1.3.29}$$

Multiplying with λ_2 and dividing with $e^{\lambda_2 t}$ gives

$$\lambda_2 N_2 = U_2 + U_1 - \frac{\lambda_2 U_1 e^{-\lambda_1 t}}{\lambda_2 - \lambda_1} - (U_2 + U_1)e^{-\lambda_2 t} + \frac{\lambda_2 U_1}{\lambda_2 - \lambda_1}e^{-\lambda_2 t} \tag{1.3.30}$$

The production rate of radionuclides is given by

$$U = \dot{\Phi}\sigma N_0 \tag{1.3.31}$$

where $\dot{\Phi}$ is the neutron fluence rate, σ is the reaction cross section and N_0 is the number of target nuclei at $t=0$.

Combining Eq. (1.3.30) with (1.3.31) gives

$$A_2 = \lambda_2 N_2 = \dot{\Phi} N_0 \left[(\sigma_2 + \sigma_1)(1 - e^{-\lambda_2 t}) + \frac{\lambda_2 \sigma_1}{\lambda_2 - \lambda_1}(e^{-\lambda_2 t} - e^{-\lambda_1 t}) \right] \tag{1.3.32}$$

When $t \rightarrow \infty$ then

$$A_2(\infty) = \dot{\Phi} N_0 \left[(\sigma_1 + \sigma_2) \right] \tag{1.3.33}$$

Data:
$\sigma_1 = 1.20 \cdot 10^{-27}\,\mathrm{m^2}$ (reaction cross section for producing $^{104m}\mathrm{Rh}$)
$\sigma_2 = 14.0 \cdot 10^{-27}\,\mathrm{m^2}$ (reaction cross section for producing $^{104}\mathrm{Rh}$)
$\lambda_1 = \ln 2/(4.40 \cdot 60)\,\mathrm{s^{-1}}$ (decay constant for $^{104m}\mathrm{Rh}$)
$\lambda_2 = \ln 2/42\,\mathrm{s^{-1}}$ (decay constant for $^{104}\mathrm{Rh}$)
$t_1 = 60\,\mathrm{s}$ (irradiation time)
$t_2 = 600\,\mathrm{s}$ (irradiation time)

Data inserted in Eq. (1.3.32) gives for $t=60$ s

$$\begin{aligned} A_2 = \dot{\Phi} N_0 [& (14.0 \cdot 10^{-27} + 1.2 \cdot 10^{-27})(1 - e^{-\ln 2 \cdot 60/42}) \\ & + \frac{\ln 2/42 \cdot 1.2 \cdot 10^{-27}}{\ln 2/42 - \ln 2/(4.40 \cdot 60)}(e^{-\ln 2 \cdot 60/42} - e^{-\ln 2 \cdot 60/(4.4 \cdot 60)})] \\ & = 8.864 \cdot 10^{-27}\,\dot{\Phi} N_0 \end{aligned}$$

Data inserted in Eq. (1.3.33) gives

$$A_2(\infty) = 15.2 \cdot 10^{-27}\,\dot{\Phi} N_0$$

This gives

$A_2(60)=0.58A_2(\infty)$

Data for $t=600$ s gives

$A_2(600)=0.98A_2(\infty)$

Answer: The ^{104}Rh activity is 58% of the saturation activity after 60 s and 98% of the saturation activity after 600 s.

Solution exercise 1.10.
The change in the number of active ^{169}Yb-nuclei is given by the equations

$$\frac{\mathrm{d}N}{\mathrm{d}t} = U(t) - \lambda N \tag{1.3.34}$$

$$U(t) = \frac{mN_A\dot{\Phi}\sigma}{m_a}e^{-\dot{\Phi}\sigma t} \tag{1.3.35}$$

where $U(t)$ is the production rate of radioactive nuclides, m is the mass of target (kg), N_A is Avogadro's number, $\dot{\Phi}$ is the neutron fluence rate, σ is the activation cross section, m_a is the atomic mass and t is the irradiation time.

The equations are solved by multiplying with the integrating factor $e^{\lambda t}$

$$e^{\lambda t}(\frac{\mathrm{d}N}{\mathrm{d}t} + \lambda N) = U(t)e^{\lambda t} \tag{1.3.36}$$

$$\mathrm{d}\left(\frac{Ne^{\lambda t}}{\mathrm{d}t}\right) = U(t)e^{\lambda t} = \frac{mN_A\dot{\Phi}\sigma}{m_a}e^{-\dot{\Phi}\sigma t}e^{\lambda t} \tag{1.3.37}$$

Integration gives

$$Ne^{\lambda t} = \frac{mN_A\dot{\Phi}\sigma}{m_a(\lambda - \dot{\Phi}\sigma)}e^{t(\lambda-\dot{\Phi}\sigma)} + C \tag{1.3.38}$$

If t=0 then N=0

$$C = -\frac{mN_A\dot{\Phi}\sigma}{m_a(\lambda - \dot{\Phi}\sigma)} \tag{1.3.39}$$

$$N = \frac{mN_A\dot{\Phi}\sigma}{m_a(\lambda - \dot{\Phi}\sigma)}\left(e^{-\dot{\Phi}\sigma t} - e^{-\lambda t}\right) \tag{1.3.40}$$

The activity is given by $A = \lambda N$. The maximal activity is obtained when $(\mathrm{d}A/\mathrm{d}t) = 0$

$$-\dot{\Phi}\sigma e^{-\dot{\Phi}\sigma t} + \lambda e^{-\lambda t} = 0 \tag{1.3.41}$$

$$\frac{\lambda}{\dot{\Phi}\sigma} = e^{(\lambda - \dot{\Phi}\sigma)t} \tag{1.3.42}$$

$$\Rightarrow t = \frac{\ln\left(\frac{\lambda}{\dot{\Phi}\sigma}\right)}{\lambda - \dot{\Phi}\sigma} \tag{1.3.43}$$

Data:

$\lambda = \ln2/32\ \text{d}^{-1} = 2.507 \cdot 10^{-7}\ \text{s}^{-1}$ (decay constant for ^{169}Yb)
$\dot{\Phi} = 1.00 \cdot 10^{18}\ \text{m}^{-2}\text{s}^{-1}$ (neutron fluence rate)
$\sigma = 1.10 \cdot 10^{-24}\ \text{m}^2$ (reaction cross section)

Data inserted in Eq. (1.3.43) gives

$t = 1.741 \cdot 10^6\ \text{s} = 20.15\ \text{d}$

Answer: The maximal activity of ^{169}Yb is obtained after 20.2 days.

Solution exercise 1.11.
The activity of ^{99}Mo on Monday 8:00 is $A_{Mo}(0)$.

The activity of ^{99}Mo on Thursday 8:00, $A_{Mo}(t_{Mo})$, is

$$A_{Mo}(t_{Mo}) = A_{Mo}(0)e^{-\lambda_{Mo}t_{Mo}}.$$

The generator is then eluted again at a time t after the first elution. The activity of the ^{99m}Tc is measured at time t_1. The activity of ^{99m}Tc at time t_1 is obtained from the relation

$$A_{Tc} = \frac{f\lambda_{Tc}A_{Mo}(0)e^{-\lambda_{Mo}t_{Mo}}}{\lambda_{Tc} - \lambda_{Mo}}\left(e^{-\lambda_{Mo}t} - e^{-\lambda_{Tc}t}\right)e^{-(t_1-t)\lambda_{Tc}} \tag{1.3.44}$$

where
$t_{Mo} = 72\,\text{h}$ (time from Monday 8:00 to Thursday 8:00)
$t_1 = 8.0\,\text{h}$ (time from first elution at 8:00 to when measuring the activity at 16:00)
$A_{Mo}(0) = 12.0\,\text{GBq}$ (activity of ^{99}Mo Monday morning 8:00)
$A_{Tc} = 2.10\,\text{GBq}$ (activity of ^{99m}Tc at Thursday 16:00)
$f = 0.876$ (fraction of ^{99}Mo disintegrations that will give ^{99m}Tc activity)
$\lambda_{Mo} = \ln2/66.0\,\text{h}^{-1}$ (decay constant for ^{99}Mo)
$\lambda_{Tc} = \ln2/6.02\,\text{h}^{-1}$ (decay constant for ^{99m}Tc)

Data inserted in Eq. (1.3.44) gives

$$2.10 = \frac{0.876 \cdot (\ln2/6.02) \cdot 12 \cdot e^{-(\ln2/6.02)\cdot72}}{\ln2/6.02 - \ln2/66.0}$$

$$\times \left(e^{-(\ln 2/66.0)\cdot t} - e^{-(\ln 2/6.02)\cdot t} \right) e^{-(8.0-t)\cdot \ln 2/6.02}$$

$t = 6.49\,\text{h}$

Answer: The elution was made at 14:30 h (t is calculated from 8:00).

Solution exercise 1.12.

When the reactor is shut down there are ^{149}Pm and ^{149}Nd radionuclides in the reactor. Both neodymium and promethium will end up in samarium through the reaction

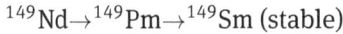

^{149}Nd→^{149}Pm→^{149}Sm (stable)

Samarium is thus obtained in two ways, either form promethium directly or from neodymium through promethium.

a) The number of ^{149}Sm nuclides that are obtained from the ^{149}Pm radionuclides directly is given by

$$N_{1,\text{Sm}} = \int_0^T A_{\text{Pm}} e^{-\lambda_{\text{Pm}} t}\, dt = \frac{A_{\text{Pm}}}{\lambda_{\text{Pm}}} \left(1 - e^{-\lambda_{\text{Pm}} T} \right) \tag{1.3.45}$$

b) The number of ^{149}Sm nuclides that are obtained from ^{149}Nd radionuclides is obtained from the equations

$$A_{\text{Pm}} = \frac{\lambda_{\text{Pm}} A_{\text{Nd}}}{\lambda_{\text{Nd}} - \lambda_{\text{Pm}}} \left(e^{-\lambda_{\text{Pm}} t} - e^{-\lambda_{\text{Nd}} t} \right) \tag{1.3.46}$$

and

$$N_{2,\text{Sm}} = \int_0^T A_{\text{Pm}}\, dt = \frac{\lambda_{\text{Pm}} A_{\text{Nd}}}{\lambda_{\text{Nd}} - \lambda_{\text{Pm}}} \left[-\frac{e^{\lambda_{\text{Pm}} t}}{\lambda_{\text{Pm}}} + \frac{e^{\lambda_{\text{Nd}} t}}{\lambda_{\text{Nd}}} \right]_0^T \tag{1.3.47}$$

$$N_{2,\text{Sm}} = \frac{\lambda_{\text{Pm}} A_{\text{Nd}}}{\lambda_{\text{Nd}} - \lambda_{\text{Pm}}} \left[\frac{e^{-\lambda_{\text{Nd}} T}}{\lambda_{\text{Nd}}} - \frac{e^{-\lambda_{\text{Pm}} T}}{\lambda_{\text{Pm}}} - \frac{1}{\lambda_{\text{Nd}}} + \frac{1}{\lambda_{\text{Pm}}} \right] \tag{1.3.48}$$

The mass is obtained from the relation

$$m = \frac{m_{\text{a}} N}{N_{\text{A}}} \tag{1.3.49}$$

where
$A_{\text{Pm}} = 1000$ TBq (activity of ^{149}Pm when the reactor is shut down)
$A_{\text{Nd}} = 300$ TBq (activity of ^{149}Nd when the reactor is shut down)
$T = 48$ h (time after reactor shut down)
$\lambda_{\text{Nd}} = \ln 2/1.7\,\text{h}^{-1}$ (decay constant for ^{149}Nd)

λ_{Pm} =ln2/47 h^{-1} (decay constant for ^{149}Pm)
m_a =149 (atomic mass)
N_A =6.022·10^{26} kmol^{-1} (Avogadro's number)

Data inserted in Eq. (1.3.45) and Eq. (1.3.48) gives

$$N_{1,Sm} = \frac{1000 \cdot 10^{12} \cdot 3600}{\ln 2/47}\left(1 - e^{-(\ln 2/47)\cdot 48}\right)$$

and

$$N_{2,Sm} = \frac{300 \cdot 10^{12}(\ln 2/47) \cdot 3600}{\ln 2/1.7 - \ln 2/47}$$
$$\times \left[\frac{e^{-(\ln 2/1.7)48}}{\ln 2/1.7} - \frac{e^{-(\ln 2/47)48}}{\ln 2/47} - \frac{1}{\ln 2/1.7} + \frac{1}{\ln 2/47}\right]$$

a) $N_{1,Sm}$= 1.2384·10^{20} nuclei
b) $N_{2,Sm}$= 1.295·10^{18} nuclei

In total there are 1.251·10^{20} nuclei. This gives the mass m= 3.095·10^{-5} kg.

Answer: The mass of ^{149}Sm after 48 h is 31 mg.

2 Interaction of Ionizing Radiation with Matter

2.1 Definitions and Relations

The definitions in the text below refer mainly to ICRU Report 85: Fundamentals Quantities and Units for Radiation (ICRU, 2011).

2.1.1 Radiometric Quantities

Radiation can be divided into two main types:

Ionizing radiation, which can ionize matter and produce ions. This is possible if the particle energy is higher than the ionization potential. As the ionization potential varies with atomic number, the threshold value for a radiation being ionizing will vary. However, this is often a minor problem in the determination of the absorbed dose as this cut-off energy is typically small. In radiobiological applications a threshold value of 10 eV is often used.

Non-ionizing radiation, which doesn't have sufficient energy to ionize and can only give rise to excitations of the atoms and molecules. These energy transfers are small and will normally have a small biological effect. This type of radiation is not included in this text.

Ionizing radiation can further be divided into two classes:

Charged particles as electrons and positrons (light charged particles) as well as protons, α-particles, and other heavy ions (heavy charged particles). Previously this type of radiation was called "directly ionizing" radiation as the particle deposits its energy directly to emission of orbital electrons through Coulomb collisions.

Uncharged particles as photons and neutrons. This type of radiation was previously called "indirectly ionizing" radiation as it deposits the energy by first emitting charged particles which then ionize.

The use of the concepts of "directly" and "indirectly ionizing radiation" should be abandoned as they are not formally correct.

Fluence

Fluence Φ is defined as

$$\Phi = \frac{\mathrm{d}N}{\mathrm{d}a} \; \text{Unit} : \mathrm{m}^{-2} \tag{2.1.1}$$

where $\mathrm{d}N$ is the number of particles incident on a sphere of cross-sectional area $\mathrm{d}a$.

The distribution, Φ_E, is defined as

$$\Phi_E = \frac{\mathrm{d}\Phi}{\mathrm{d}E} \tag{2.1.2}$$

where $d\Phi$ is the fluence of particles of energy between E and $E + dE$.

Energy fluence Ψ is defined as

$$\Psi = \frac{dR}{da} \text{ Unit} : J\,m^{-2} \tag{2.1.3}$$

where dR is the radiant energy incident on a sphere of cross-sectional area da.

The distribution, Ψ_E, is defined as

$$\Psi_E = \frac{d\Psi}{dE} \tag{2.1.4}$$

where $d\Psi$ is the energy fluence of particles of energy between E and $E + dE$.

The relationship between the two distributions is given by

$$\Psi_E = E\Phi_E \tag{2.1.5}$$

The corresponding time dependent quantities are called *fluence rate* and *energy fluence rate* and are defined as

$$\dot{\Phi} = \frac{d\Phi}{dt} \text{ Unit} : m^{-2}\,s^{-1} \tag{2.1.6}$$

and

$$\dot{\Psi} = \frac{d\Psi}{dt} \text{ Unit} : W\,m^{-2} \tag{2.1.7}$$

The *particle radiance*, $\dot{\Phi}_\Omega$ is defined as

$$\dot{\Phi}_\Omega = \frac{d\dot{\Phi}}{d\Omega} \text{ Unit} : m^{-2}\,s^{-1}\,sr^{-1} \tag{2.1.8}$$

where $d\dot{\Phi}$ is the fluence rate of particles propagating within a solid angle $d\Omega$ around a specified direction.

The *energy radiance*, $\dot{\Psi}_\Omega$ is defined as

$$\dot{\Psi}_\Omega = \frac{d\dot{\Psi}}{d\Omega} \text{ Unit} : W\,m^{-2}\,sr^{-1} \tag{2.1.9}$$

where $d\dot{\Psi}$ is the energy fluence rate of particles propagating within a solid angle $d\Omega$ around a specified direction.

The *solid angle* is defined as $d\Omega = \sin\theta\,d\theta\,d\phi$ and illustrated in Fig. 2.1

2.1.2 Definition of Interaction Coefficients

Cross section
The *cross section* of a target entity, σ, for a particular interaction produced by incident

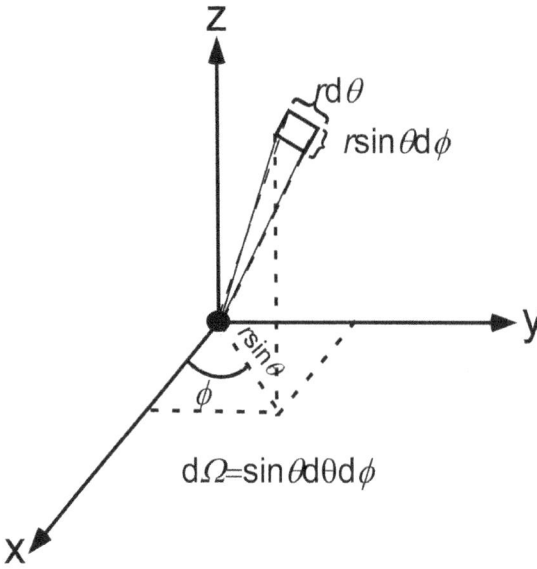

Figure 2.1: Illustration of the solid angle. Adapted after Attix (1986).

charged or uncharged particles is defined as the quotient of N by Φ, where N is mean number of such interactions per target for a particle fluence Φ. Note that the cross section is an area and with larger area the probability to hit a target is larger and the cross section is also larger. This area is not equal to the "geometrical area" of the target (atom, nucleus etc), but varies with incident particle and energy for the same target.

$$\sigma = \frac{N}{\Phi} \text{ Unit : m}^2 \tag{2.1.10}$$

An important differential cross section is the differential cross section per solid angle $d\sigma/d\Omega$, i.e. the probability per solid angle that an incident particle is scattered into a solid angle $d\Omega$. The total cross section σ is then obtained by

$$\sigma = \int \frac{d\sigma}{d\Omega} \, d\Omega \tag{2.1.11}$$

This differential cross section per solid angle shall be separated from the differential cross section per angle, $d\sigma/d\Theta$. The relation between $d\sigma/d\Omega$ and $d\sigma/d\Theta$ is given by

$$\frac{d\sigma}{d\Theta} = \frac{d\sigma}{d\Omega} 2\pi\sin\Theta. \tag{2.1.12}$$

This cross section goes to zero when Θ goes to zero as $\sin 0 = 0$.

Mass attenuation coefficient

The *mass attenuation coefficient*, μ/ρ is the quotient of dN/N by $\rho \, dl$, where dN/N is the fraction of uncharged particles that interacts when passing a distance dl in a material with density ρ.

$$\frac{\mu}{\rho} = \frac{1}{\rho \, dl} \frac{dN}{N} \quad \text{Unit}: \text{m}^2 \, \text{kg}^{-1} \tag{2.1.13}$$

The advantage of using μ/ρ is that it is independent of the density, ρ, of the material. It is the coefficient that is tabulated in most tables. When calculating the attenuation in a material then the thickness should be expressed in $\text{kg} \, \text{m}^{-2}$. It is also possible to use the thickness of a material expressed in m, in which case the linear attenuation coefficient, μ, is used. The reciprocal of μ is called the *mean free path* or sometimes the *relaxation length*.

The mass attenuation coefficient is obtained by adding the cross sections of the different interaction types.

$$\frac{\mu}{\rho} = \frac{N_A}{M} \sum_J \sigma_J \tag{2.1.14}$$

where σ_J is the component cross section relating to interaction type J, N_A is Avogadro's number and M is the molar mass.

The types of interaction for photons normally included in tabulated values of μ/ρ are photoelectric effect, coherent scattering, incoherent scattering, and pair production in the electron and the nuclear field. For some reasons the nuclear cross sections are normally not included, even if they can contribute to some per cent of the total cross section for some energies.

μ/ρ may thus be expressed as

$$\frac{\mu}{\rho} = \frac{\tau}{\rho} + \frac{\sigma_{\text{coh}}}{\rho} + \frac{\sigma_{\text{incoh}}}{\rho} + \frac{\kappa_n}{\rho} + \frac{\kappa_e}{\rho} \tag{2.1.15}$$

where τ/ρ is the mass cross section for photoelectric effect, σ_{coh}/ρ for the coherent (Rayleigh) scattering, $\sigma_{\text{incoh}}/\rho$ for the incoherent scattering (Compton scattering), κ_n/ρ for pair production in the nuclear field, and κ_e/ρ is the mass cross section in the electron field.

The mass attenuation coefficient of a compound is obtained by treating the compound as consisting of independent atoms, neglecting any molecular binding energies. Thus

$$\left(\frac{\mu}{\rho}\right)_{\text{comp}} = w_1 \left(\frac{\mu}{\rho}\right)_1 + w_2 \left(\frac{\mu}{\rho}\right)_2 + .. + w_j \left(\frac{\mu}{\rho}\right)_j \tag{2.1.16}$$

where w is the mass fraction of each atomic material in the compound. This relation is sometimes called the Bragg relation. This is a good approximation except for very low photon energies.

Mass energy transfer coefficient
The *mass energy transfer coefficient*, μ_{tr}/ρ, is the quotient of the mean energy that is transferred to kinetic energy of charged particles by interaction of incident uncharged

particles and the incident radiant energy, when passing a distance dl in a material with density ρ.

$$\frac{\mu_{tr}}{\rho} = \frac{1}{\rho\,dl}\frac{dR_{tr}}{R} \quad \text{Unit} : m^2\,kg^{-1} \tag{2.1.17}$$

The mass energy transfer coefficient is related to the mass attenuation coefficient by

$$\frac{\mu_{tr}}{\rho} = \frac{\mu}{\rho}f \tag{2.1.18}$$

where

$$f = \frac{\sum f_J \sigma_J}{\sum \sigma_J} \tag{2.1.19}$$

where f_J is the average fraction of the incident particle energy that is transferred to kinetic energy of charged particles in an interaction of type J.

The different contributions from the separate interaction types are with some approximations given by the relations below.

<u>Photoelectric effect</u>. The electrons emitted in the photoelectric effect are the photoelectrons and the Auger electrons including the Coster-Cronig electrons. The energy not obtained as kinetic energy of the electrons is obtained as energy of fluorescence x rays. It is more practical to calculate the emitted photon energy and deduct it from the primary photon energy. The mass energy transfer coefficient τ_{tr}/ρ for the photoelectric effect can be written as

$$\tau_{tr}/\rho = (\tau/\rho)\left[1 - \frac{p_K \omega_K h\nu_K}{h\nu} - \frac{(1 - p_K)p_L \omega_L h\nu_L}{h\nu} - \dots\right] \tag{2.1.20}$$

where p_K and p_L are the probability that the photoelectric effect will occur in the K and the L-shell respectively. ω_K and ω_L are the fluorescence yields. $h\nu_K$ and $h\nu_L$ are the mean energies of the x rays from the K- and L-shell respectively.

<u>Incoherent scattering</u>. To calculate the mean energy of the electrons emitted in incoherent scattering, it is possible in the first approximation to use the Klein-Nishina relation to obtain the electron energy distribution. This mean energy is independent of the atomic number and data for this and the corresponding cross-sections are tabulated in Table 2.1.

<u>Pair production</u>. In pair production the electron and positron will obtain the photon energy, apart from the energy needed to produce the electron-positron pair, $2m_e c^2$. Thus

$$T_- + T_+ = h\nu - 2m_e c^2 \tag{2.1.21}$$

and

$$\kappa_{tr}/\rho = (\kappa/\rho)\frac{h\nu - 2m_e c^2}{h\nu} \tag{2.1.22}$$

Mass energy absorption coefficient

The *mass energy absorption coefficient* is the product of the mass energy transfer coefficient μ_{tr}/ρ and (1-g)

$$\mu_{en}/\rho = \mu_{tr}/\rho(1-g) \text{ Unit} : \text{m}^2 \text{ kg}^{-1} \tag{2.1.23}$$

where g is the fraction of energy liberated by charged particles which is lost in radiative processes as bremsstrahlung and annihilation in flight. μ_{en}/ρ is an important coefficient when calculating the absorbed dose as it gives the energy absorbed in collisions of charged particles with electrons, and thus to ionization in the medium. For low photon energies and low atomic numbers the numerical difference between μ_{en}/ρ and μ_{tr}/ρ is small as the contribution to bremsstrahlung is small. With increasing energy and atomic number the difference between μ_{tr}/ρ and μ_{en}/ρ increases.

Stopping power

The *stopping power*, S/ρ, is the quotient of dE by ρdl, where dE is the energy lost by a charged particle when passing a distance dl in a material with density ρ.

$$\frac{S}{\rho} = \frac{1}{\rho} \frac{dE}{dl} \text{ Unit} : \text{J m}^2 \text{ kg}^{-1} \tag{2.1.24}$$

S/ρ is in tables often expressed in MeV m^2 kg^{-1} as the particle energies often are given in MeV.

The mass stopping power can be divided into a sum of independent components.

$$\frac{S}{\rho} = \frac{1}{\rho}\left(\frac{dE}{dl}\right)_{el} + \left(\frac{dE}{dl}\right)_{rad} + \left(\frac{dE}{dl}\right)_{nuc} = \left(\frac{S}{\rho}\right)_{el} + \left(\frac{S}{\rho}\right)_{rad} + \left(\frac{S}{\rho}\right)_{nuc} \tag{2.1.25}$$

where

$(S/\rho)_{el}$ is the mass collision (electronic) stopping power due to collisions with electrons. This is the quantity to be used in absorbed dose calculations.

$(S/\rho)_{rad}$ is the mass radiative stopping power due to bremsstrahlung emitted in the electric field of atomic nuclei or atomic electrons. In medical radiation physics $(S/\rho)_{rad}$ is of interest only for electrons and positrons as the probability for bremsstrahlung is inversely proportional to the particle mass in square, and thus can be neglected for protons and heavier charged particles in the energy ranges used in radiotherapy.

$(S/\rho)_{nuc}$ is the mass nuclear stopping power due to elastic Coulomb collisions in which recoil energy is imparted to atoms. This part is often small and neglected, in particular for electrons.

Critical energy

The *critical energy* is the energy when $(S/\rho)_{el}$ equals $(S/\rho)_{rad}$. For lower energies $(S/\rho)_{el}$ dominates and for higher energies $(S/\rho)_{rad}$. The *critical energy* varies with atomic number and is around 10 MeV for lead and 100 MeV for water. An empirical

Table 2.1: Klein-Nishina cross-sections expressed in 10^{-21} m^2/electron and mean energies for the scattered photons ($\bar{h\nu}$) and electrons ($\bar{T}$). σ is the total cross-section and $\sigma_s = \sigma(\bar{h\nu}/h\nu)$ and $\sigma_a = \sigma(\bar{T}/h\nu)$.

Energy/MeV	$_e\sigma$	$_e\sigma_s$	$_e\sigma_a$	$\bar{h\nu}$/MeV	$\bar{T}$/MeV
0.010	6.405	6.285	0.120	0.0098	0.00019
0.015	6.290	6.116	0.174	0.0146	0.00041
0.020	6.180	5.957	0.223	0.0193	0.00072
0.030	5.976	5.664	0.311	0.0284	0.00156
0.040	5.788	5.401	0.387	0.0373	0.00268
0.050	5.615	5.162	0.453	0.0460	0.00404
0.060	5.456	4.946	0.511	0.0544	0.00562
0.080	5.173	4.567	0.606	0.0706	0.00937
0.10	4.928	4.248	0.680	0.0862	0.0138
0.15	4.436	3.631	0.805	0.123	0.0273
0.20	4.065	3.186	0.879	0.157	0.0433
0.30	3.535	2.582	0.953	0.219	0.0809
0.40	3.167	2.186	0.981	0.276	0.124
0.50	2.892	1.904	0.987	0.329	0.171
0.60	2.675	1.692	0.983	0.379	0.221
0.80	2.395	1.389	0.960	0.473	0.327
1.00	2.112	1.183	0.929	0.560	0.440
1.50	1.716	0.867	0.849	0.758	0.742
2.00	1.464	0.687	0.777	0.938	1.062
3.00	1.151	0.586	0.665	1.268	1.732
4.00	0.960	0.377	0.583	1.572	2.428
5.00	0.829	0.308	0.520	1.860	3.140
6.00	0.732	0.261	0.472	2.136	3.863
8.00	0.599	0.199	0.400	2.662	5.338
10.0	0.510	0.161	0.349	3.164	6.836
15.0	0.377	0.109	0.268	4.349	10.65
20.0	0.303	0.083	0.220	5.469	14.53
30.0	0.220	0.056	0.164	7.586	22.41
40.0	0.175	0.042	0.133	9.598	30.40
50.0	0.146	0.034	0.112	11.54	38.46
60.0	0.125	0.028	0.097	13.43	46.57
80.0	0.099	0.021	0.078	17.08	69.92
100.0	0.082	0.017	0.065	20.62	79.38

value for the critical energy is

$$T_{\text{crit}} = \frac{800}{Z + 1.2} \text{ MeV} \tag{2.1.26}$$

where Z is the atomic number.

Linear energy transfer

The *linear energy transfer* or the *restricted stopping power*, L_Δ, is the quotient of dE_Δ

by dl, where dE_Δ is the energy lost by a charged particle due to electronic collisions traversing a distance dl, minus the sum of the kinetic energies of all electrons released with kinetic energies in excess of Δ.

$$L_\Delta = \frac{dE_\Delta}{dl} \quad \text{Unit} : \text{J}\,\text{m}^{-1} \tag{2.1.27}$$

L_Δ is sometimes expressed in keV/μm. The Δ-value is often expressed in eV. Then L_{100} should be understood to be the linear energy transfer for an energy cutoff of 100 eV. L_Δ is also sometimes written as S_Δ. L_Δ (S_Δ) is an important quantity both in dosimetry and in radiobiology.

Mass scattering power
The *mass scattering power* is defined as the increase in mean square angle of scattering ($d\bar{\theta}^2$) per unit mass thickness ($\rho\,dl$) in analogy with the mass stopping power.

$$\frac{T}{\rho} = \frac{1}{\rho}\frac{d\bar{\theta}^2}{dl} \quad \text{Unit} : \text{radian}^2\,\text{m}^2\,\text{kg}^{-1} \tag{2.1.28}$$

When a beam of electrons is impinging on a medium, the mean square scattering angle at small depths may be given by the relation

$$\bar{\theta}^2(x) = \bar{\theta}^2(0) + \frac{T}{\rho}\rho x. \tag{2.1.29}$$

2.1.3 Interaction Processes

The information in this section is mainly focussed on the energies obtained by the secondary particles in the interaction processes, as these are of interest in dosimetry. For more information on interaction processes consult the relevant textbooks, e.g. Podgorsak (Podgorsak, 2010).

Photoelectric effect
In a photoelectric process the photon is totally absorbed by the atom and the energy is transferred to an atomic electron, often a K-electron, which is then is emitted (released from its orbit). The probability for a photoelectric effect approximately varies with energy and atomic number according to the relationship

$$\tau/\rho \propto \frac{Z^4}{(h\nu)^3} \tag{2.1.30}$$

Close to the energy of the electron binding energies, there is a large discontinuous jump in the cross section value, because if the photon energy is just below e.g. the binding energy of the K-shell, then no K-electrons can be expelled, but with an energy just above the binding energy it is possible. Tables of photon cross sections always have two lines corresponding to the binding energy. One gives the value of the cross section just below the binding energy and the other gives the value of the cross section just above the binding energy.

The energy of the photoelectron is given by the relation

$$T_{K,L,M,..} = h\nu - B_{K,L,M,..} \tag{2.1.31}$$

where $B_{K,L,M,..}$ is the binding energy of the K, L, M,... electrons respectively. The electron vacancy produced in a photoelectric effect is followed by the emission of fluorescence x rays or Auger and Coster-Cronig electrons when an electron from an outer shell fills the vacancy. For further information see Chap. 1: *Radiation sources and Radioactive decay*.

Incoherent scattering

In incoherent scattering, the photon is scattered by an electron. The photon energy is divided between the scattered photon and the emitted electron. When the electron is assumed to be free and at rest the process is often called Compton scattering.

The cross section for Compton scattering per electron is independent of the atomic number, as the electron is supposed to be free, and thus the mass cross section is proportional to the number of electrons per mass unit. This quantity is rather constant with atomic number, except for hydrogen, and thus σ/ρ varies slowly with atomic number. Taking the binding energy of the electron into consideration, the cross section for low photon energies and high atomic number is however considerably decreased.

In the calculations of incoherent scattering in this compilation, the electron is assumed to be free and at rest and thus it fulfills the requirements for the Compton and the Klein-Nishina relations.

The Klein-Nishina relation gives the probability that a photon is scattered into a solid angle $d\Omega$ around the angle Θ. See Fig. 2.2.

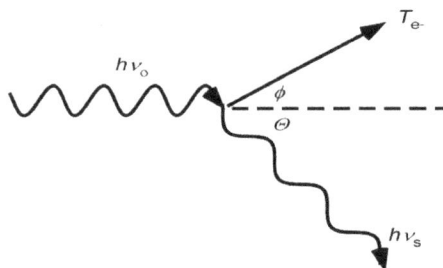

Figure 2.2: Incoherent scattering of a photon.

The Klein-Nishina differential cross section is given by

$$\frac{d_e \sigma^{KN}(\theta)}{d\Omega} = \frac{r_e^2}{2} \frac{1}{[1 + \alpha(1 - \cos\theta)]^2} \times \left[1 + \cos^2\theta + \frac{\alpha^2(1 - \cos\theta)^2}{1 + \alpha(1 - \cos\theta)}\right] \frac{m^2/\text{electron}}{\text{steradian}}$$

(2.1.32)

α is the incident photon energy expressed in the rest mass energy of an electron.

$$\alpha = \frac{h\nu}{m_e c^2}$$

(2.1.33)

r_e is the "classical electron radius" which may be written as

$$r_e = \frac{1}{4\pi\epsilon_o} \frac{e^2}{m_e c^2} = 2.818 \cdot 10^{-15} \text{ m}$$

(2.1.34)

The energy of the scattered photon is given by

$$h\nu_s = \frac{h\nu}{1 + \frac{h\nu}{m_e c^2}(1 - \cos\theta)}$$

(2.1.35)

The energy of the emitted electron is given by

$$T_e = h\nu - h\nu_s$$

(2.1.36)

The minimum energy of the scattered photon is obtained when it is scattered in 180°. The energy is then given by

$$h\nu_{s,\min} = \frac{h\nu}{1 + 2\alpha}$$

(2.1.37)

When the primary photon energy tends to ∞ the minimum energy tends to $m_e c^2/2$, or 0.256 MeV.

Correspondingly the electron energy is at its maximum when the electron is emitted in the forward direction. This energy is given by

$$T_{e,\max} = \frac{2\alpha}{1 + 2\alpha} h\nu$$

(2.1.38)

Note that the scattered photon has a minimum non-zero energy and thus the emitted electron has a maximum energy different from the primary photon energy. At high photon energies the minimum photon energy is close to 0.25 MeV and thus the maximum electron energy is close to $h\nu - 0.25$ MeV.

In absorbed dose calculations the mean energy of the Compton electron is often of interest. The relative mean energy varies with the incident photon energy and for low photon energies most of it is transferred to the photon and the relative electron energy is low. With increasing photon energy the relative amount of energy transferred to the electron increases and is close to complete transfer at very high photon energies. See Table 2.1

Pair production

For high photon energies the photon can interact through pair production, either in the field of a nucleus or in the field of an electron. The photon is totally absorbed and its energy is used to produce an electron-positron pair.

The cross section for pair production increases after the threshold value, first nearly logarithmically with energy, but reaches a saturation value for very high photon energies. The cross section for pair production in the nuclear field is always larger than the cross section in the electron field and as such the ratio is close to the atomic number in the first approximation.

The energy threshold for pair production in a nuclear field is $2m_e c^2 = 1.022$ MeV. The energy threshold for pair production in an electron field (sometimes called triplet production) is $4m_e c^2 = 2.044$ MeV. For energies larger than the threshold energy, the excess energy is divided between the electron-positron pair. As the positron is an anti-particle it will recombine with an electron and two annihilation photons will be produced. If the positron has no kinetic energy left when it is annihilated, then the energy of each annihilation photon is 0.511 MeV each, and they are emitted in opposite directions. This is the most common situation, but there is a small probability that the positron is annihilated when having kinetic energy. This is called "annihilation in flight" in which case the annihilation photons will have a higher energy and not move in opposite directions.

Transmission of photons

When a narrow mono-energetic photon beam with a fluence Φ_o and energy $h\nu$ passes through a material with the mass attenuation coefficient μ/ρ m^2 kg^{-1} and thickness ρx m^2 kg^{-1} , the fluence $\Phi_{\rho x}$ after transmission is given by

$$\Phi_{\rho x} = \Phi_o e^{-(\mu/\rho)(\rho x)} \tag{2.1.39}$$

This relation holds for narrow beams that are well collimated both before and after the absorbing material and thus only transmitted primary photons are included. In most situations there are also contributions of secondary photons, mainly incoherent scattered photons, but also annihilation photons and fluorescence x rays. If the coherent scattered photons, with the same energy as the primary ones, shall be considered as secondary or not, can depend on the geometry. The contributions from secondary photons are often included by multiplying the transmitted fluence with a *buildup factor, B*, which is the ratio of the total transmitted fluence to the primary fluence.

$$\Phi_{\rho x} = B\Phi_o e^{-(\mu/\rho)(\rho x)} \tag{2.1.40}$$

The buildup factor will not be treated in this chapter but in Chap. 6 *Radiation protection and health physics*.

Absorption of electrons

As electrons interact more or less continuously and lose energy, the transmission of electrons can not be treated as for photons where there is a low probability for interaction and a photon can pass through thick materials without interacting. However, there are some relations that can be useful in radiation protection situations where the accuracy is not that important. β-particles are emitted isotropically and with a continuous energy distribution. This has shown to result in a transmission more or less exponential and thus the transmission can be written as

$$\Phi_{\rho x} = \Phi_o e^{-(\beta/\rho)(\rho x)} \tag{2.1.41}$$

where β/ρ is the mass absorption coefficient for β-particles. The value of β/ρ varies with energy and atomic number of the absorber. The values have been obtained experimentally. One expression for low atomic numbers is

$$\beta/\rho = \frac{3.5Z}{AT_\beta^{1.14}} \ \text{m}^2 \, \text{kg}^{-1} \tag{2.1.42}$$

For high atomic numbers instead the following expression is often used

$$\beta/\rho = \frac{0.77Z^{0.31}}{T_\beta^{1.14}} \ \text{m}^2 \, \text{kg}^{-1} \tag{2.1.43}$$

T_β is the maximum β-particle energy in MeV, Z is the atomic number and A is the atomic mass. These expressions are empirical and approximate and should only be used in the first part of the absorption curve.

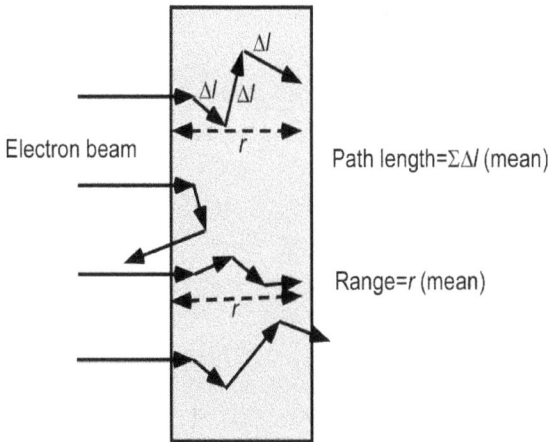

Figure 2.3: Passage of charged particles through a material. The concepts Range and Path length are indicated.

Range. Range for a charged particle may be described in different ways. One way is to follow the path of the particle and calculate the total *path length*. Another way is to calculate the thickness of a material thick enough to totally stop the charged particles, see Fig. 2.3, which is often of more interest. For heavy charged particles which are losing their energy more or less continuously in many small steps and are not significantly scattered, the values for the different range descriptions are very close. However, for electrons, with a large scattering probability and a possibility to loose a lot of energy in one collision there may be a large difference depending on which definition is used. Also in a beam of electrons the ranges will be varying extensively between the electrons and normally one has to calculate a mean value. It is thus important to include which range definition is used when a range value is presented. The value often tabulated is r_{csda}, the continuous slowing down approximation. This is defined as

$$r_{csda} = \int_0^{T_o} \frac{dT}{S_{tot}(T)} \tag{2.1.44}$$

This is thus based on the assumption that the electrons are losing energy continuously and gives a path length.

Empirical expressions for particle ranges are often used in radiation protection. Two of these expressions are given below:

$$R_1 = 4.12 T_\beta^{(1.265-0.095 \ln T_\beta)} \text{ kg m}^{-2} \tag{2.1.45}$$

$$R_2 = 5.30 T_\beta - 1.06 \text{ kg m}^{-2} \tag{2.1.46}$$

where T_β is the maximal β-energy (MeV).

These relations are mainly obtained for aluminum, but may be used for other low atomic number materials. The expressions show that the ranges in g cm^{-2} are approximately equal to half of the electron energy in MeV. In water the range in cm is just half of the electron energy. A 10 MeV electron beam has thus approximately a range of 5 cm in water, a 20 MeV beam 10 cm, and so on.

Electron energy. The electrons lose energy when passing through a medium and there will be an energy distribution that broadens with depth. This distribution is not totally symmetrical and this implies that the mean energy and the most probable energy differ. In dosimetry the mean energy is often of most interest.

The mean energy of the electrons at a depth x may be approximated by the relation

$$\bar{T}(x) = \bar{T}_0 - \frac{dT_0}{\rho dl} \rho x \tag{2.1.47}$$

where $\bar{T}(x)$ is the mean kinetic energy at depth x, $\bar{T}_0$ is the mean kinetic energy at surface and $\frac{dT_0}{\rho dl}$ is the mass stopping power at surface.

This approximation assumes that the stopping power does not change with depth and thus energy, which is not correct. Brahme (Brahme, 1975) has proposed an approximation (Eq. (2.1.48)) for variation of the mean energy with depth, which holds well for the first half of the electron range.

$$\bar{T}(x) = \bar{T}_0 - \frac{dT}{\rho dl}\frac{1 - e^{-\rho x \epsilon_{rad,0}}}{\epsilon_{rad,0}} \tag{2.1.48}$$

where

$$\epsilon_{rad,0} = \frac{1}{\bar{T}_0}\frac{dT_{rad,0}}{\rho dl} \tag{2.1.49}$$

An approximation for the mean energy proposed by Harder (Harder, 1965) assumes a linear decrease of the mean energy with depth. The mean energy is expressed as

$$\bar{T}(x) = \bar{T}_0(1 - x/r_p) \tag{2.1.50}$$

where r_p is the electron range.

However, this expression holds better for the most probable energy and should not be used when accurate calculations are necessary. In dosimetry today most data of electron energies and thus stopping power values are based on Monte Carlo calculations and these data should be used if possible.

2.2 Exercises in Interaction of Ionizing Radiation

2.2.1 Charged Particles

Exercise 2.1. What is the maximal δ-particle energy that can be obtained for a proton, an electron and positron respectively, if their kinetic energy is 10 MeV?

Exercise 2.2. Electrons with a mean energy of 20 MeV impinge perpendicularly on a water phantom. Estimate the mean energy at the depth of 3.0 cm and 8.0 cm, using different approximations?

Exercise 2.3. Electrons with an energy of 2.3 MeV impinge perpendicularly on a disk of a) PMMA b) lead. Calculate the mean square scattering angle at the thicknesses 2.0 resp. 5.0 kg m^{-2}. Discuss any approximations you have made.

Exercise 2.4. When treating patients with electrons, the electron beam is obtained from an electron accelerator. This electron beam is narrow when it leaves the accelerator, with a diameter of some millimeters. However, for a stationary beam, the beam profile at the patient should be large and uniform. This is obtained by placing scattering foils in the beam, which scatter out the electrons and thus broaden it. At the same time as the electrons are scattered, they will lose energy and produce bremsstrahlung. These are unwanted effects.

An electron beam has an energy of 10 MeV. Which material should be chosen as a scattering foil to obtain as small energy loss and production of bremsstrahlung as

Table 2.2: Klein-Nishina differential cross-section for the number of photons scattered per unit solid angle in direction θ, $d(_e\sigma)/d\Omega$, in 10^{-26} cm^2/steradian per electron.

Energy MeV	θ 1°	5°	10°	20°	30°	40°	50°	60°	70°	80°	90°	120°	150°	180°
0.01	7.940	7.910	7.817	7.459	6.912	6.243	5.534	4.868	4.324	3.962	3.821	4.687	6.472	7.360
0.02	7.940	7.909	7.814	7.448	6.891	6.209	5.488	4.812	4.260	3.890	3.738	4.539	6.218	7.051
0.03	7.940	7.907	7.807	7.424	6.741	6.132	5.385	4.687	4.117	3.729	3.554	4.216	5.679	6.399
0.04	7.940	7.906	7.803	7.407	6.805	6.077	5.313	4.602	4.020	3.622	3.434	4.014	5.346	6.000
0.05	7.940	7.905	7.798	7.389	6.770	6.024	5.243	4.519	3.928	3.522	3.323	3.831	5.049	5.643
0.06	7.940	7.904	7.794	7.372	6.736	5.971	5.175	4.440	3.841	3.427	3.219	3.664	4.781	5.324
0.08	7.940	7.902	7.784	7.338	6.668	5.868	5.043	4.288	3.677	3.253	3.031	3.371	4.319	4.777
0.10	7.939	7.899	7.775	7.304	6.600	5.768	4.918	4.147	3.527	3.097	2.866	3.124	3.936	4.328
0.20	7.939	7.887	7.729	7.138	6.283	5.313	4.371	3.560	2.937	2.515	2.277	2.313	2.730	2.928
0.30	7.939	7.876	7.684	6.979	5.992	4.922	3.933	3.124	2.530	2.140	1.919	1.866	2.100	2.213
0.40	7.938	7.864	7.639	6.825	5.724	4.581	3.575	2.789	2.235	1.880	1.679	1.583	1.716	1.783
0.50	7.938	7.852	7.595	6.679	5.481	4.293	3.301	2.569	2.081	1.785	1.626	1.562	1.661	1.708
0.60	7.937	7.841	7.551	6.537	5.251	4.022	3.029	2.313	1.837	1.542	1.373	1.236	1.269	1.292
0.80	7.936	7.818	7.464	6.269	4.847	3.584	2.636	1.991	1.580	1.328	1.180	1.026	1.014	1.017
1.00	7.935	7.795	7.380	6.021	4.497	3.233	2.339	1.759	1.397	1.176	1.042	0.882	0.847	0.842
2.00	7.931	7.681	6.980	5.011	3.299	2.190	1.538	1.156	0.923	0.775	0.677	0.527	0.471	0.456
3.00	7.926	7.571	6.617	4.278	2.608	1.679	1.173	0.882	0.703	0.586	0.506	0.378	0.328	0.314
4.00	7.921	7.463	6.287	3.728	2.163	1.373	0.957	0.718	0.570	0.472	0.405	0.295	0.251	0.239
5.00	7.916	7.358	5.986	3.301	1.854	1.167	0.812	0.607	0.480	0.396	0.337	0.242	0.204	0.193
6.00	7.912	7.256	5.710	2.962	1.626	1.018	0.706	0.527	0.415	0.341	0.289	0.205	0.172	0.162
8.00	7.902	7.058	5.223	2.460	1.312	0.815	0.563	0.417	0.327	0.267	0.225	0.157	0.130	0.123
10.0	7.893	6.870	4.808	2.107	1.104	0.682	0.468	0.346	0.269	0.219	0.184	0.128	0.105	0.099
20.0	7.846	6.051	3.422	1.245	0.627	0.379	0.256	0.186	0.144	0.116	0.097	0.066	0.053	0.050
30.0	7.800	5.394	2.652	0.894	0.441	0.263	0.176	0.128	0.098	0.079	0.065	0.044	0.036	0.034
40.0	7.754	4.858	2.167	0.701	0.341	0.202	0.134	0.097	0.074	0.060	0.049	0.033	0.027	0.025
50.0	7.709	4.414	1.836	0.578	0.278	0.164	0.109	0.078	0.060	0.048	0.040	0.027	0.022	0.020
60.0	7.665	4.041	1.595	0.492	0.235	0.138	0.091	0.066	0.050	0.040	0.033	0.022	0.018	0.017
80.0	7.577	3.454	1.268	0.379	0.179	0.104	0.069	0.050	0.038	0.030	0.025	0.017	0.014	0.013
100.0	7.490	3.014	1.055	0.309	0.145	0.084	0.056	0.040	0.030	0.024	0.020	0.013	0.011	0.010

possible, for a certain mean scattering angle? Motivate your statement considering the variation with atomic number of the different interaction processes.

Exercise 2.5. When treating patients with photons, the photons are often produced by electrons from an electron accelerator hitting a target, whereby bremsstrahlung is obtained.

A beam of 20 MeV electrons hits targets of aluminum or tungsten (same thickness corresponding to the electron range). The photon energy fluence was measured at the central axis at a distance of 100 cm from the target, and a higher energy fluence was obtained when using aluminum, even if the radiation yield is higher for tungsten. How can this be explained?

Exercise 2.6. A narrow beam of 1000 MeV protons, impinges on a PMMA-cone at its peak along the cone axis. Calculate the opening angle of the cone if the emitted Cerenkov radiation, after reflection, towards the wall of the cone passes through the bottom of the cone and parallel with its axis.

Exercise 2.7. A spherical ionization chamber with a radius of 0.30 m has a ^{32}P-source positioned at its center. What air pressure is needed to completely absorb the β-particles in the air volume at a temperature of 293 K? Estimate also the ionization current if the activity of the source is 40.0 kBq. $\bar{W}_{air}$=33.97 eV. $\bar{W}_{air}$ is the mean energy needed to produce an ion pair in air.

Exercise 2.8. An arm with a total thickness of 12 cm is irradiated with 150 MeV protons. The bone thickness is 4.0 cm and the total soft muscle tissue thickness is 8.0 (4.0 + 4.0) cm (see Fig. 2.4). Is the proton energy enough for the proton to pass through the arm, and if so, what is the energy of the protons after passing through the arm? ρ_{bone}=1.8·10^3 kg m^{-3}, $\rho_{soft\ tissue}$=1.0·10^3 kg m^{-3}.

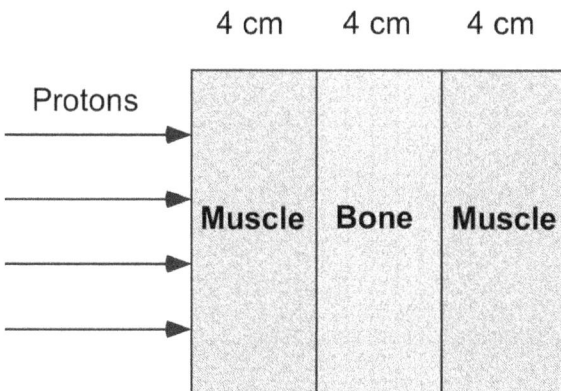

Figure 2.4: Simulation of the arm in exercise 2.8.

2.2.2 Photons

Exercise 2.9. There are three "absorption" coefficients defined for photons, the mass attenuation coefficient (μ/ρ), the mass energy transfer coefficient (μ_{tr}/ρ), and the mass energy absorption coefficient (μ_{en}/ρ). The ratios $(\mu/\rho)/(\mu_{tr}/\rho)$ and $(\mu_{en}/\rho)/(\mu_{tr}/\rho)$ for lead are plotted in Fig. 2.5. Explain the variation of the ratios with energy.

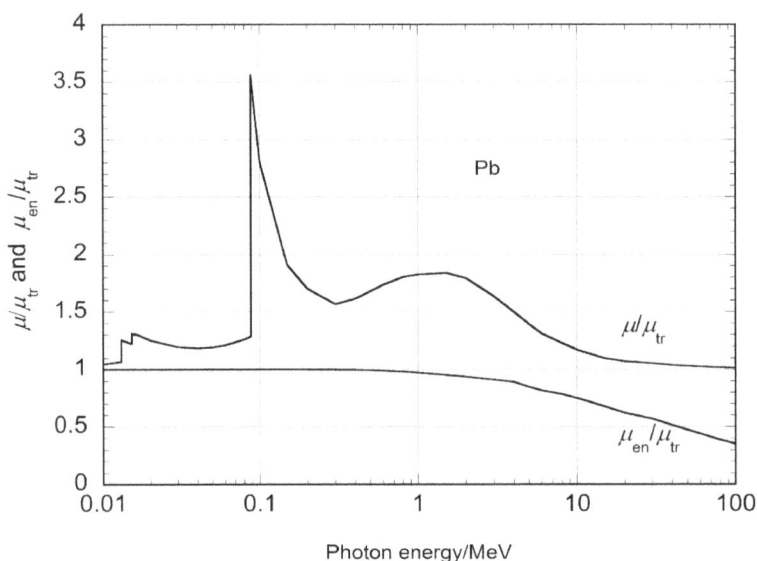

Figure 2.5: Variation with photon energy of the ratios mass attenuation coefficient and mass energy absorption coefficient to mass energy transfer coefficient in lead.

Exercise 2.10. In order to obtain information of the treated volume in radiotherapy with high energy photons, the use of produced positrons that may be measured with a positron camera, has been proposed. It is possible to measure positrons emitted only during the irradiation and positrons emitted also even after the treatment is finished. From which type of photon interaction processes are these positrons obtained? Which are the minimum photon energies for these two different processes? Discuss the energy distribution of the positrons in the two processes.

Exercise 2.11. Different materials were irradiated with diagnostic x rays with a maximal energy of 140 keV (mean energy 55 keV). The secondary photons were measured at an angle of 90° to the primary beam, both with and without an absorber made of PMMA, with a thickness of 10 mm, between the scattering material and the detector. Fig. 2.6 shows the ratio between the fluence without and with the PMMA absorber as a function of the atomic number of the irradiated material. Discuss the ratio with knowledge of the interaction processes in the irradiated material.

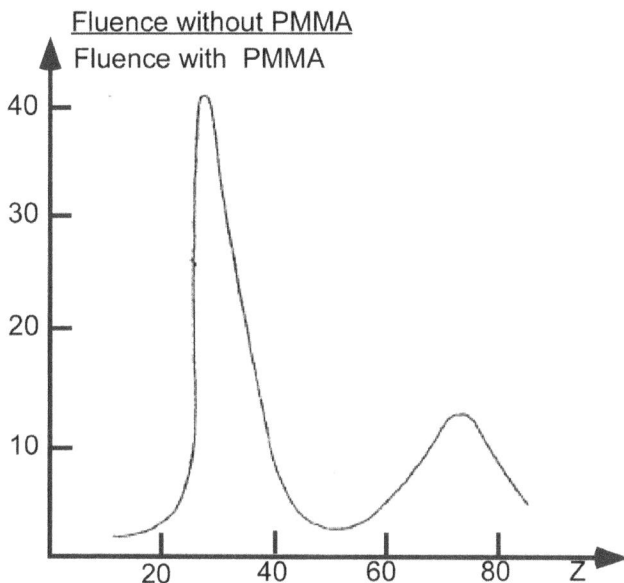

Figure 2.6: Ratio of fluence of secondary photons, without and with an absorber of PMMA, as a function different atomic numbers of scattering material.

Exercise 2.12. Lead is a good photon absorber. Sometimes the fluorescence x rays produced in connection with the photoelectric effect may be a problem. To reduce this effect the lead absorber is surrounded with iron to absorb the x rays. What thickness of iron is needed (in mm) to reduce the fluorescence x-ray fluence a factor of 10.0? Only the K x rays need to be considered. Assume a narrow beam geometry.

Exercise 2.13. In diagnostic radiological investigations it is important to decrease the number of photons scattered in the patient that can reach the detector. These scattered photons will deteriorate the quality of the image as they have another direction than the primary photons. To reduce this component a grid is often placed in front of the detector. This grid is supposed to absorb the scattered photons having another direction than the primary ones. The grid (see Fig. 2.7) can be made of parallel slits of lead with a supporting material, with a high transmission for x rays in between. In the figure the material is aluminum. Assume you have a grid with dimensions as in the figure. Calculate the ratio between the transmission of primary and scattered photons for the situation where the scattered photons have the highest probability for transmission, i.e. when the attenuation in the lead slits is a small as possible. The energy is supposed to be 50 keV for the primary photons and 40 keV for the scattered photons.

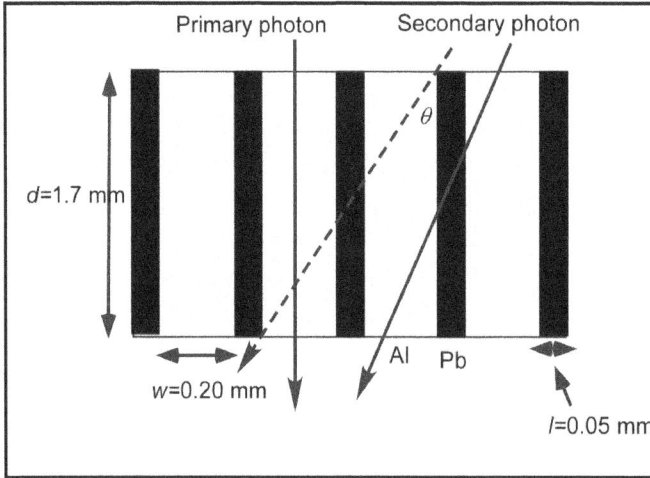

Figure 2.7: Illustration of passage of photons through a grid for diagnostic x-rays.

Exercise 2.14. Determine the smallest photon energy where the photon can interact through incoherent scattering, and an equal large energy transfer to the scattered photon and the emitted electron is possible. In which angles are then the electron and scattered photon emitted?

Exercise 2.15. An experiment was performed as shown in Fig. 2.8. 100 keV photons hit a small rod of aluminum. A NaI-scintillator is positioned at different angles in relation to the impinging photon beam. In front of the NaI-scintillator a thin filter of lead is positioned that reduces the photon fluence. When the angle increases from zero, the counting rate decreases. At a certain angle there is, however, a large increase in the counting rate. Then the counting rate again decreases with increasing angle. Calculate the angle where this increase occurs. Calculate also the thickness of the lead-filter if the the photon fluence was increased with a factor of 10.

Exercise 2.16. Calculate the energy of a Compton scattered photon with the primary energy 1.17 MeV if it is scattered in the angle $\pi/2$ radians a) in one scattering b) in two equal scatterings c) in three equal scatterings d) in an infinite number of equal small scatterings.

Exercise 2.17. A narrow collimated beam of 2.04 MeV photons passes through a very thin layer of lead. The secondary electrons that are emitted in the angle 20° are

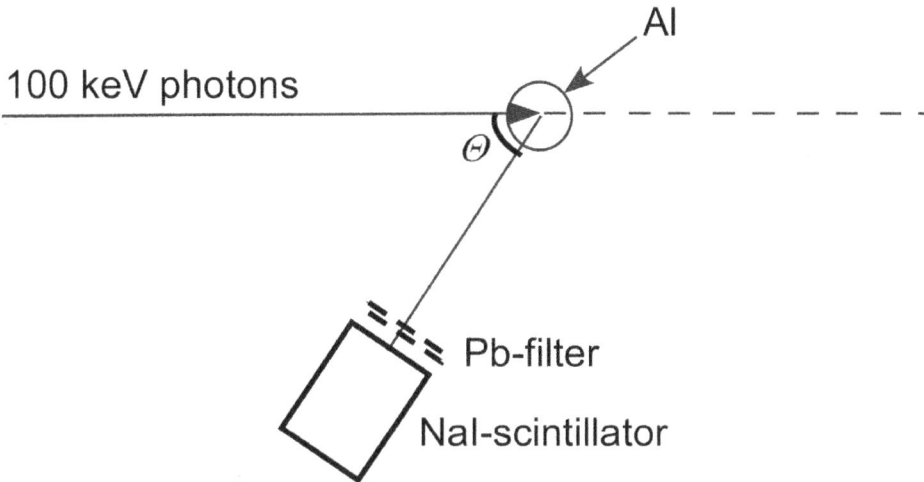

Figure 2.8: Illustration of irradiation geometry in exercise 2.21.

observed. Which is the energy in this angle for the
a) photoelectrons?
b) Compton electrons?
c) pair production electrons?

Exercise 2.18. Calculate the mass energy transfer coefficient for 2.0 MeV photons in lead with knowledge of the cross sections for the different interaction processes.

Exercise 2.19. To flatten a photon beam from an accelerator, a beam flattening filter is placed in the beam. This filter is however a source for secondary particles. Estimate the fluence from secondary photons 10 cm from the central axis (B) as compared to the fluence from primary photons at the central axis (A). See Fig. 2.9. In this calculation the filter may be assumed to have a uniform thickness of 7.0 mm Pb and a diameter of 50 mm. The accelerator produces photons with a mean energy of 2.0 MeV. When calculating the number of photon interactions one can assume that all secondary photons are produced along the central axis.

Exercise 2.20. In radiotherapy, lead blocks are often placed in the beam to limit the absorbed dose to critical organs that not should be irradiated. These blocks are often placed on a thin disk placed in the beam. The photons will then produce electrons and positrons (leptons) through interactions in the disk. These leptons may reach the patient. The contamination of these leptons should be as low as possible. Compare the lepton contamination directly behind disks of PMMA and lead, both with the thickness 15 kgm^{-2}. The parallel photon fluence at the surface of the disk is $1.0 \cdot 10^{15}$

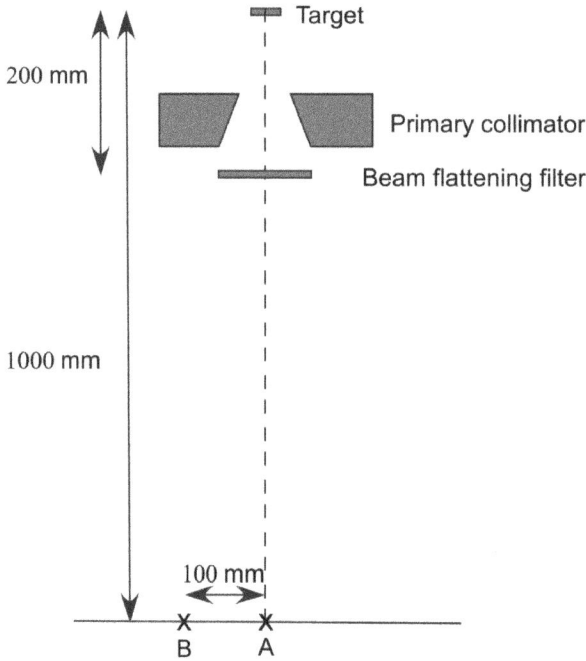

Figure 2.9: Sketch of a treatment head with a beam flattening filter to be used in the calculations.

m^{-2}. The electrons are supposed to be absorbed exponentially in the disk and the absorption coefficient, β, is obtained from the equations below. T_{max} is the maximal energy of the leptons in MeV. The radiation source is a linear accelerator with the acceleration energy 21 MeV. The photon energy distribution may be approximated with a mean energy of 7 MeV.

$$\beta_{PMMA} = \frac{3.5Z}{AT_{max}^{1.14}} \text{ m}^2 \text{ kg}^{-1}$$

$$\beta_{Pb} = \frac{0.77Z^{0.31}}{T_{max}^{1.14}} \text{ m}^2 \text{ kg}^{-1}$$

Exercise 2.21. A small sphere of lead is irradiated with 100 keV photons. The secondary photon fluence rate at a distance of 200 mm is determined. A sphere of graphite with the same mass is placed at the same place as the lead sphere and irradiated with the same primary photon fluence. At what angle is the same energy of the secondary photons obtained as with the lead sphere? Calculate the ratio of the fluence rate of the secondary photons at a distance of 200 mm in this angle, with the graphite sphere and with the lead sphere. Neglect the self-absorption in the spheres that may be regarded as point sources. In the lead case only K x rays may be included in the calculations and in the graphite case only Compton scattered photons. The energy of the K x rays may be approximated with the energy of the $K_{\alpha II}$ x rays.

Exercise 2.22. The fluence of the bremsstrahlung from a target irradiated by high energy electrons has an angular distribution that can be expressed as

$$\Phi(\Theta) = \frac{\Phi(0)}{1 + \frac{(T\Theta)^{1.4}}{1.73^{1.4}}} \qquad (2.2.1)$$

where T is the electron energy in MeV and Θ is the angle in radians.

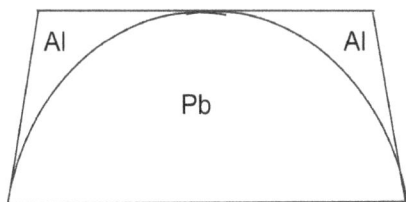

Figure 2.10: A beam flattening filter with two different atomic numbers to flatten both the fluence and the energy.

The aim is to have a flattened uniform beam at the phantom surface. The field size at the phantom surface is 40x40 cm^2. Distance target - phantom = 100 cm. The beam flattening filter has a constant thickness, but is made of lead at the center and aluminum at the edges. See Fig. 2.10. Calculate the ratio between the thicknesses of lead and aluminum to have a flattened beam at a distance of 10 cm from the central axis. The mean energy of the photons is 7 MeV at the axis and 6 MeV at the field edge and a linear variation of the energy with the distance from the central axis may be assumed. The electrons from the accelerator have an energy of 21 MeV.

Exercise 2.23. A narrow beam of 50 MeV photons with a radius 3.0 mm is impinging on a disk of lead with a thickness of 10 mm. The fluence rate is $3.0 \cdot 10^{10}$ m^2 s^{-1}. Behind the disk, at a distance of 50 mm and an angle of 45°, a thin lead foil (thickness: 0.10 mm, area 10.0 mm^2) is situated. At an angle of 45° to this foil, a detector is placed at a distance of 50 mm (See Fig. 2.11). The detector area is 500 mm^2. Calculate how many photons in the energy intervals 80±10 keV and 400±50 keV that will hit the detector per hour. Simplify the calculations by neglecting the attenuation of the photons in the disks.

Exercise 2.24. To measure x-ray spectra directly is often difficult as the count rate is very high and pile-up effects in the detectors are obtained. Instead it is possible to measure the radiation scattered in 90°. Using the Compton equation it is then possible to recalculate the measured energies to the energies before scattering. It

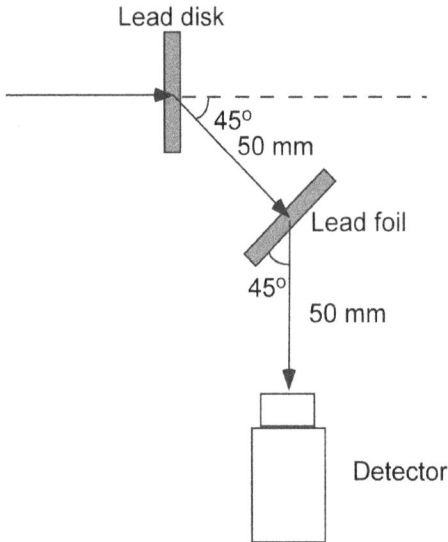

Figure 2.11: Irradiation geometry in exercise 2.23.

is however important that the photons are just scattered once, which is why a very thin scattering material is used. In this experiment a very narrow cylinder made of a material with a low atomic number, to decrease the contribution from the photoelectric effect, is used. The number of scattered photons is then very small and it important to shield the detector from direct radiation from the radiation source according to Fig. 2.12. The scattering cylinder has a diameter of 1.0 mm and is made of aluminum. The length of the irradiated cylinder is 12 mm. Neglect the self-absorption in the cylinder. Calculate the thickness of the lead shield needed, in order to have a fluence rate of the transmitted photons that is less than 2.0% of the scattered photons, using narrow beam approximation. Only the contribution from incoherently scattered photons has to be included in the calculations. Assume that the isotropic radiation source emits monoenergetic photons with an energy of 80 keV.

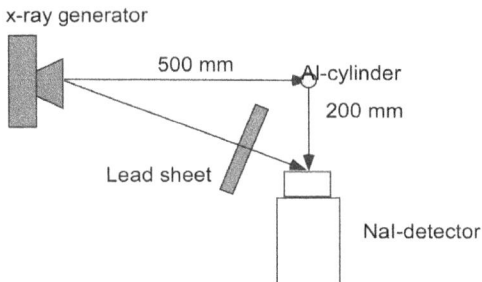

Figure 2.12: Illustration of the irradiation geometry in exercise 2.24.

Exercise 2.25. One way to measure the content of lead in a human being is to measure the fluorescence radiation that is obtained when the body is irradiated with photons with an energy higher than the binding energy of the K-electrons. Lead is mainly attached to the bone tissue. To investigate the efficiency of such an investigation the following experiment is performed. An arm is simulated with a tube filled with water. The tube diameter is 100 mm. At the center of the tube a cylinder of aluminum, with a diameter of 30 mm, simulating the bone tissue, is placed. At the center of this cylinder there is a hole with the diameter 1.0 mm. See Fig. 2.13.

In this hole a cylinder of lead is positioned. The tube is irradiated with 100 keV photons and the photon fluence at the center of the cylinder without any attenuation is $4.5 \cdot 10^{12}$ m^{-2}. The diameter of the radiation beam is 30 mm at the center of the cylinder. Calculate the photon fluence in the energy interval 80-90 keV at a point 500 mm from the cylinder center, at an angle of 120° to the direction of the primary photons. The secondary photons may be assumed to come from the center of the lead cylinder.

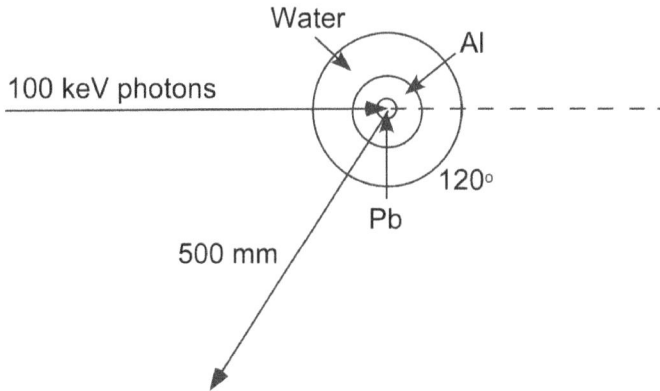

Figure 2.13: Simulation of the arm in exercise 2.25.

2.3 Solutions in Interaction of Ionizing Radiation

2.3.1 Charged Particles

Solution exercise 2.1.

a) The maximal energy transfer is obtained when the incoming particle hits the electron straight on, i.e. the electron is emitted in the forward direction (see Fig 2.14).

Proton   Electron

Before collision

 M, v_1 m_e

After collision

 M, v_2 m_e, v_3

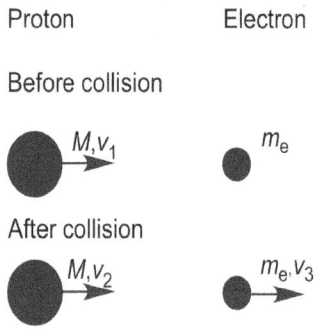

Figure 2.14: Collision of a proton with an electron in a straight on collision.

The relation for the kinetic energy gives

$$\frac{Mv_1^2}{2} = \frac{Mv_2^2}{2} + \frac{m_e v_3^2}{2} \qquad (2.3.1)$$

The momentum relation gives

$$Mv_1 = Mv_2 + m_e v_3 \qquad (2.3.2)$$

Reorganizing and dividing the equations gives

$$\frac{M(v_1^2 - v_2^2)}{M(v_1 - v_2)} = \frac{m_e v_3^2}{m_e v_3} \qquad (2.3.3)$$

and

$$v_1 + v_2 = v_3 \qquad (2.3.4)$$

If M >> m_e, the energy transfer is small and $v_1 \approx v_2$, giving $v_3 \approx 2v_1$

This results in that the maximal energy transfer is

$$\Delta E_{max} = \frac{m_e v_3^2}{2} \approx \frac{4 m_e v_1^2}{2} = 2 m_e v_1^2 \qquad (2.3.5)$$

Inserting $v_1^2 = 2T/M$, where T is the kinetic energy of the incoming particle gives

$$\Delta E_{max} = \frac{2 m_e 2T}{M} = \frac{4 m_e}{M} T \qquad (2.3.6)$$

With the incoming particle being a proton, the kinetic energy 10 MeV and the particle mass $1836 m_e$, this results in

$$\Delta E_{max} = \frac{4 \cdot 1}{1836} \cdot 10 \cdot 10^3 = 21.8 \, \text{keV}$$

b) When the incoming particle is an electron the total kinetic energy may be transferred to another free electron at rest. However, as after the collision there will then be two electrons sharing the energy, and as in most cases, it does not matter which electron is which, it is common to regard the electron with the highest energy after the collision to be the primary one. Thus the maximal energy transfer is half of the incoming energy; in this case 5.0 MeV.

c) If the incoming particle is a positron, it is possible to separate the particles and the total energy transfer is 10 MeV.

Answer: The maximal energy transfer is a) 21.8 keV (proton), b) 5 MeV (electron) and c) 10 MeV (positron).

Solution exercise 2.2.

The mean energy of the electrons at a depth x may be approximated by the relation

$$\bar{T}(x) = \bar{T}_0 - \frac{1}{\rho}\frac{dT_0}{dl}\rho x \tag{2.3.7}$$

where $\bar{T}(x)$ is the mean kinetic energy at depth x, $\bar{T}_0$ is the mean kinetic energy at surface and $\frac{1}{\rho}\frac{dT_0}{dl}$ is the mass stopping power at surface.

Data:
$\frac{1}{\rho}\frac{dT_0}{dl}$ (20 MeV)=0.2455 MeV m^2 kg^{-1}
a) ρx=30 kg m^{-2}, b) ρx=80 kg m^{-2}.

Assume that the density of water is $1.0\cdot10^3$ kg m^{-3}.

Data inserted gives

$\bar{T}$(3 cm)=20-0.2455·30=12.6 MeV
$\bar{T}$(8 cm)=20-0.2455·80=0.36 MeV

This approximation assumes that the stopping power does not change with depth and thus energy, which is not correct. Brahme (Brahme, 1975) has proposed an approximation (Eq. (2.3.8)) for variation of the mean energy with depth, which holds well for the first half of the electron range.

$$\bar{T}(x) = \bar{T}_0 - \frac{dT}{\rho\,dl}\frac{1 - e^{-\rho x \epsilon_{rad,0}}}{\epsilon_{rad,0}} \tag{2.3.8}$$

where

$$\epsilon_{rad,0} = \frac{1}{\bar{T}_0}\frac{dT_{rad,0}}{\rho\,dl} \tag{2.3.9}$$

$\epsilon_{rad,0}$(20MeV)=0.04086/20=2.043·10^{-3} m^2 kg^{-1}

Data inserted gives

$$\bar{T}(3) = 20 - \frac{0.2455(1 - e^{-30 \cdot 2.043 \cdot 10^{-3}})}{2.043 \cdot 10^{-3}} = 12.86 \text{ MeV}$$

and

$$\bar{T}(8) = 20 - \frac{0.2455(1 - e^{-80 \cdot 2.043 \cdot 10^{-3}})}{2.043 \cdot 10^{-3}} = 1.88 \text{ MeV}$$

An approximation for the mean energy proposed by Harder (Harder, 1965) assumes a linear decrease of the mean energy with depth. The mean energy is expressed as

$$\bar{T}(x) = \bar{T}_0(1 - x/r_0) \tag{2.3.10}$$

where r_0 is the electron range.

IAEA TRS 381 (IAEA, 1997) presents Monte Carlo calculated ranges and calculations for 20 MeV electrons give a range of 9.96 cm. This inserted in Eq. (2.3.10) gives

$\bar{T}(3.0)=14$ MeV and $\bar{T}(8.0)=4.0$ MeV.

This relation normally overestimates the mean energy at small depths and underestimates the energy at depths close to the range.

To obtain a correct result of the mean energy Monte Carlo calculations should be performed. IAEA TRS 381 gives data for the variation of mean energy with depth. The mean energy is 12.6 MeV at a depth of 3.0 cm, and 2.94 MeV at a depth of 8.0 cm. This can then be considered to be close to the correct mean energy.

This indicates that for small depths the first approximation above is quite accurate, but for larger depths the approximation proposed by Brahme is closer to the correct value.

Answer: The mean energy at a thickness of 3.0 cm is 12.6-14 MeV and at a thickness of 8.0 cm 0.4-4 MeV depending on the approximation.

Solution exercise 2.3.
In the first approximation the mean square scattering angle is given by the relation

$$\bar{\theta}^2(x) = \bar{\theta}^2(0) + \frac{T}{\rho}\rho x \tag{2.3.11}$$

where $\bar{\theta}^2(0)$ is the mean square scattering angle of the incoming beam and T/ρ the mass scattering power.

Data:

$\bar{\theta}^2(0)=0$ (parallel perpendicularly incoming beam)

$\frac{T}{\rho}=0.0770$ radian^2 m^2 kg$^-$1 (2.3 MeV, PMMA)

$\frac{T}{\rho}=0.664$ radian^2 m^2 kg$^-$1 (2.3 MeV, Pb)

$\rho x=2.0$ kg m^{-2} or 5.0 kg m^{-2}

Data inserted gives

PMMA:

$\bar{\theta}^2(2)=0.0770\cdot2.0=0.154$ rad^2. This corresponds to $\sqrt{\bar{\theta}^2}=22.5°$

$\bar{\theta}^2(5)=0.0770\cdot5.0=0.385$ rad^2. This corresponds to $\sqrt{\bar{\theta}^2}=35.6°$

Pb

$\bar{\theta}^2(2)=0.664\cdot2.0=1.328$ rad^2. This corresponds to $\sqrt{\bar{\theta}^2}=66°$

$\bar{\theta}^2(5)=0.664\cdot5.0=3.32$ rad^2. This corresponds to $\sqrt{\bar{\theta}^2}=104°$

This approximation gives unrealistic results at large depths as the mean square scattering angle increases linearly with depth and reaches values giving angular distributions with mean angles larger than 90°. With increasing depth the mean square scattering angle first increases nearly linearly, but after some depth there will be a diffusion equilibrium, where the inscatter and outscatter compensate. The value of equilibrium is estimated to be obtained for $\bar{\theta}^2$ between 0.6 and 0.65 rad^2, if only the electrons in the forward direction are included. The approximation above also neglects the change in the electron energy with depth. This will increase the mass scattering power with depth and the angular equilibrium will be reached even faster.

Answer: PMMA: $\bar{\theta}^2(2.0)=0.154$ rad^2. $\bar{\theta}^2(5.0)=0.385$ rad^2. Pb:$\bar{\theta}^2(2.0)=1.328$ rad^2. $\bar{\theta}^2(5.0)=3.32$ rad^2.

Solution exercise 2.4.

Both the mass scattering power and the mass radiative power vary with the atomic number with Z^2. As the mean square scattering angle is proportional to the mass scattering power this means to obtain a certain angular distribution the thickness of the scattering foil is proportional to Z^2. Thus with increasing atomic number there is a decreasing thickness of the scattering foil for a given scattering angle. With a high atomic number there is also a higher mass radiative power increasing also with Z^2. This means that the scattering and radiative power compensate each other and the production of bremsstrahlung for a certain angle is independent of the atomic number. Thus the only factor that affects the energy loss is the mass collision stopping power. As the mass collision stopping power decreases with increasing atomic number, a high atomic number of the scattering foil will give the lowest energy loss.

Answer: A high atomic number material is suitable, e.g. gold.

Solution exercise 2.5.
When the electrons hit the target the electrons interact mainly through three processes:
a) Collisions with atomic electrons (giving lower energy with depth)
b) Inelastic collisions with the atomic nucleus (giving bremsstrahlung)
c) Elastic collisions with the atomic nucleus (elastic scattering)

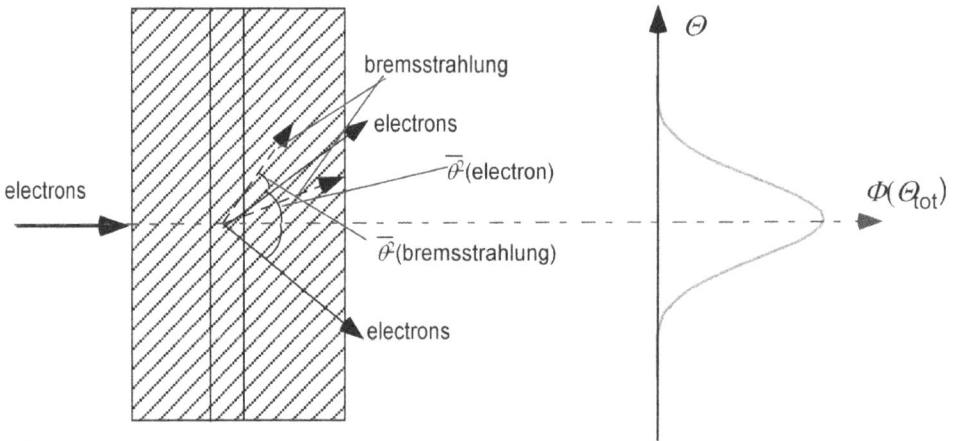

Figure 2.15: Components of the angular distribution of Bremsstrahlung photons.

The angular distribution of the bremsstrahlung beam is dependent on both the angular distribution of the electron beam and the angular distribution of the bremsstrahlung photons. The angular distribution of the bremsstrahlung photons can be described by the equation.

$$\bar{\theta}^2 = \bar{\theta}^2(1) + \bar{\theta}^2(2) \tag{2.3.12}$$

where (see Fig. 2.15) $\bar{\theta}^2$ is the total mean square scattering angle of the bremsstrahlung beam. $\bar{\theta}^2(1)$ is the mean square scattering angle of electrons at depth z and $\bar{\theta}^2(2)$ is the mean square scattering angle of the bremsstrahlung photons around an electron.

The mass radiative power is increasing with Z^2, and thus tungsten will produce more bremsstrahlung photons. However, in a low atomic number material like aluminum, the elastic scattering is low and thus $\bar{\theta}^2(1)$ is also small. This means that the electron beam is not scattered to a great extent and the bremsstrahlung angular distribution will be narrow. When measuring at the central axis, thus there will be a higher fluence for aluminum than for tungsten, even if the total emitted radiation is larger for tungsten. For a full range target the attenuation of the produced bremsstrahlung in the target itself will also be higher in a tungsten target than in an aluminum target.

Solution exercise 2.6.

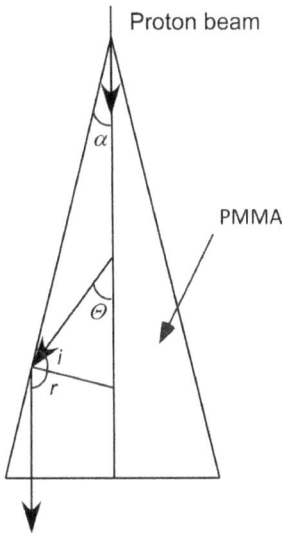

Figure 2.16: Production and transport of Cerenkov radiation in a PMMA-cone.

Cerenkov radiation is emitted at an angle θ according to the relation

$$\cos \theta = \frac{1}{\beta n} \tag{2.3.13}$$

where β is v/c (particle velocity relative the velocity of light in vacuum) and n is the index of refraction.

The particle velocity is obtained by the relation

$$\beta^2 = \frac{T(T + 2Mc^2)}{(T + Mc^2)^2} \tag{2.3.14}$$

where T is the kinetic energy of the particle and M is the particle rest mass.

Data:

$n=1.5$ (PMMA)

$T=1000$ MeV

$M c^2=938.3$ MeV (proton)

Data inserted gives

$$\beta^2 = \frac{1000(1000 + 2 \cdot 938.3)}{(1000 + 938.3)^2} = 0.766$$

This gives $\beta=0.875$.

and

$$\cos \theta = \frac{1}{0.875 \cdot 1.5} = 0.762$$

$\theta=40.4°$

The reflection angle (r) is equal to the impinging angle (i). This results (see Fig. 2.16) in $\theta = 2\alpha$. Thus $\alpha=20.2°$.

Answer: The opening angle is $20°$.

Solution exercise 2.7.

The range should be equal to the radius, r, of the ionization chamber expressed in kg m^{-2}, i.e. ρr. The density of the air is given by

$$\rho = \rho_0 \frac{p T_0}{p_0 T}$$

where p_0, p is the pressure at NTP and the pressure at measurement respectively, T_0, T is the temperature at NTP and the temperature at measurement respectively and ρ_0 is the density of air at NTP.

The range of β-particles is given by different empirical relations. Two relations often used are

$$R_1 = 4.12 T_\beta^{(1.265-0.095 \ln T_\beta)} \text{ kg m}^{-2} \tag{2.3.15}$$

$$R_2 = 5.30 T_\beta - 1.06 \text{ kg m}^{-2} \tag{2.3.16}$$

where T_β is the maximal β-energy (MeV).

These relations are mainly derived for aluminum, but may be used for other low atomic materials.

Data:

$T_\beta=1.710$ MeV (^{32}P)

p_0=101.3 kPa
T_0=273.1 K, T=293 K
ρ_0=1.293 kg m^{-3}
r=0.30 m

Data inserted in the relations for the β-particle range gives

$$R_1 = 4.12 \cdot 1.71^{(1.265-0.095\,\ln 1.71)} = 7.90\,\text{kg m}^{-2}$$

$$R_2 = 5.30 \cdot 1.71 - 1.06 = 8.00\,\text{kg m}^{-2}$$

The pressure in the chamber is obtained by setting the range of the β-particles (7.90 kg/m^2) equal to the radius of the chamber.

$$7.44 = 0.30 \cdot 1.293 \frac{p \cdot 273.1}{101.33 \cdot 293}$$

This gives p=2.21·10^3 kPa. With a range of 8.00 kg m^{-2}, p=2.24·10^3 kPa.
The ionization current is given by the equation

$$I = \frac{Af\bar{T}_\beta}{\bar{W}_{air}/e} \tag{2.3.17}$$

where
A=40·10^3 Bq (source activity)
f=1.0 (number of β-particles per decay)
$\bar{T}_\beta$=0.695·1.602 · 10^{-13} J (mean energy of the β-particles)
$\bar{W}_{air}$=33.97 eV (mean energy per produced ion pair in air)

Data inserted in Eq. (2.3.17) gives

$$I = \frac{40 \cdot 10^3 \cdot 1 \cdot 0.695 \cdot 1.602 \cdot 10^{-13}}{33.97} = 1.31 \cdot 10^{-10}\,A$$

Answer: The pressure is 2.21 or 2.24 MPa, depending on the range equation. The ionization current is 0.13 nA.

Solution exercise 2.8.
As the energy distribution of the protons is narrow and the scattering power is small, it is possible to assume that the protons travel straightforward and are nearly mono-energetic. Then it is possible to use the knowledge of residual ranges to calculate the energy at a certain depth.

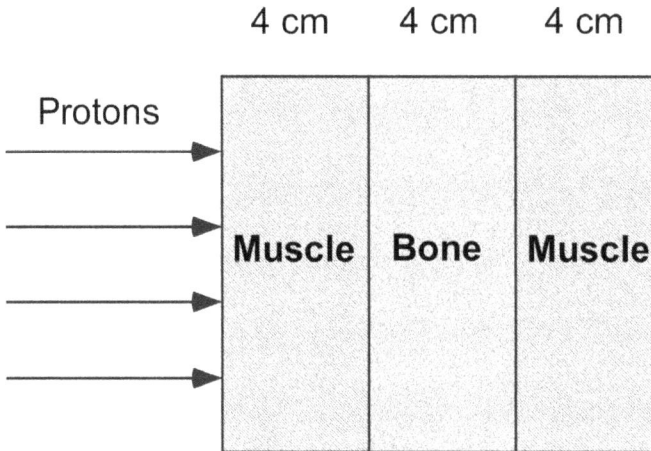

Figure 2.17: Simulation of the arm in exercise 2.8.

The range of 150 MeV protons in soft tissue (Muscle, striated ICRU) is 159.1 kg m^{-2}. The residual range in soft tissue after passing through the first part of the arm is 159.1-40=119.1 kg m^{-2}. This corresponds to an energy of 128 MeV, which is the energy of the protons when entering the bone.

The range of 128 MeV protons in bone (bone, compact, ICRU) is 128.6 kg m^{-2}.

The residual range after passing the bone is 128.6-0.04·1.8·10^3 kg/m^3=56.9 kg m^{-2}. This corresponds to an energy of 80.7 MeV.

The range of 80.7 MeV protons in soft tissue is 53.2 kg m^{-2}. The thickness of the last soft tissue part of arm is 40 kg m^{-2}. Thus the protons can pass the arm.

The residual range of the protons after passing through the arm is 53.2-40=13.2 kg m^{-2}. This corresponds to an energy of 37.2 MeV.

Answer: The protons have an energy of 37 MeV after passing through the bone.

2.3.2 Photons

Solution exercise 2.9.

The ratio (μ_{en}/ρ) to (μ_{tr}/ρ) decreases slowly with energy. The relation between (μ_{en}/ρ) and (μ_{tr}/ρ) is defined as (ICRU, 2011)

$$\mu_{en}/\rho = \mu_{tr}/\rho(1 - g)$$

where g is the part of the kinetic energy of the charged particles that is lost in radiative processes. In this example the charged particles are electrons and positrons and the main way the particles lose the energy in the form of photons is through bremsstrahlung and to a smaller part when the positrons are annihilated in flight.

With increasing energy the probability for bremsstrahlung increases as well as pair production, giving rise to more annihilation photons. This explains why the ratio decreases from a value close to unity at low energies to less than 0.5 at 100 MeV. The radiation yield at 100 MeV in Pb is around 0.76.

μ_{tr}/ρ is defined as (ICRU, 2011)

$$\frac{\mu_{tr}}{\rho} = \frac{\mu}{\rho}f$$

where

$$f = \frac{\sum f_J \sigma_J}{\sum \sigma_J}$$

and f_J is the average fraction of the incident particle energy that is transferred to kinetic energy of charged particles in an interaction type J.

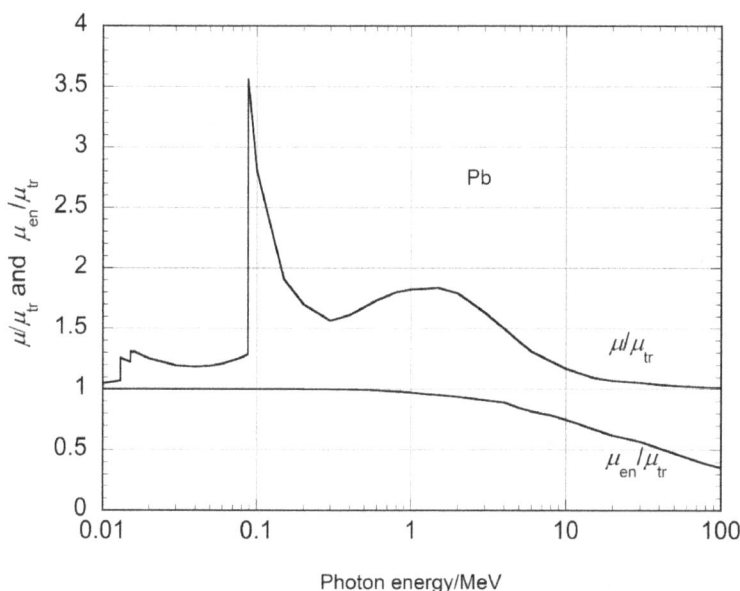

Figure 2.18: Variation with photon energy of the ratios mass attenuation coefficient and mass energy absorption coefficient to mass energy transfer coefficient in lead.

The ratio $(\mu/\rho)/(\mu_{tr}/\rho)$ has a more complex variation with energy, depending on which interaction process that dominates. At very low energies below around 100 keV the photoelectric effect dominates and below the binding energies of the L-electrons, around 15 keV, there will mainly be interactions with the M-electrons and in principle all energy will be transferred to the photoelectrons or the Auger electrons. The fine structure of the L-shells is not included in the figure. Above the L-shell binding energies, L-electrons are emitted and some of the excitation energy will be obtained as fluorescence L-x rays, resulting in a ratio slightly larger than unity. At 88 keV, the binding energy of the K-electron, there is suddenly a sharp increase in the ratio. For photons with an energy just above the binding energy, nearly no energy is left for the kinetic energy of the photoelectron and the fluorescence yield is high. This will result in a high ratio. With increasing photon energy, the photoelectrons will get more and more energy and the ratio decreases.

With higher photon energies also the probability of photoelectric effect will decrease and the incoherent scatter will be more and more important. Between 500 and 600 keV the incoherent scattering begins to dominate over the photoelectric effect and at 1.5 MeV the cross section is around ten times the cross section for photoelectric effect. At these energies the incoherent scattered electrons get slightly less than half of the photon energy resulting in a σ/σ_{tr} slightly less than two. Thus when the photoelectric effect loses influence the ratio μ/μ_{tr} increases. However, with further increasing energy the fraction of energy going to the electron in incoherent scattering is increasing and thus the ratio decreases for photon energies over around 1.5 MeV.

For energies above 45 MeV, pair production begins to dominate over incoherent scattering. Most of the photon energy is transferred to the electron-positron pair and this fraction is increasing with increasing photon energy. At 100 MeV around 99 percent of the photon energy is transferred to the electron-positron pair and the ratio is close to unity.

Solution exercise 2.10.
The positrons produced only during the irradiation are obtained through the pair production process, which is possible for photons with an energy over 1.022 MeV (pair production in the nuclear field) or 2.044 MeV (pair production in the electron field). The energy of these positrons is obtained by the relation

$$T_+ + T_- = h\nu - 2m_ec^2 \tag{2.3.18}$$

where T_+, T_- are the kinetic energies of the positron and the electron(s), $h\nu$ is the energy of the photon and m_ec^2 is the rest energy of an electron.

The positron and the electron(s) share the energy with all possibilities between zero and $h\nu - 2m_ec^2$. The energy distribution is rather uniform with the nearly same probability for all possible energies. The positrons get slightly higher energies, in particular at low photon energies, where the interactions are close to the nucleus, as the positive charge of the nucleus repels the positive charged positron and attracts the negative charged electron.

The positrons can also be obtained through (γ,n) reactions where a photon expels a neutron from the nucleus. In some situations the new produced isotope may be radioactive as e. g. in the reactions $^{12}C(\gamma,n)^{11}C$, $^{14}N(\gamma,n)^{13}N$ and $^{16}O(\gamma,n)^{15}O$, which are around in tissue. The produced radioactive nuclei emit positrons. The half lives are 20.34 min for ^{11}C, 9.96 min for ^{13}N and, 12 s for ^{15}O respectively. The energy of the photons needed for a (γ,n) reaction depends on the nuclei and is 18.7 MeV for ^{12}C, 10.55 MeV for ^{14}N, and 15.6 MeV for ^{16}O. The energies of the positrons are independent of the energy of the incoming photon and depends on the nuclear energy in the disintegration. The energy is divided between the positron and the neutrino, and there is an energy distribution of the positrons with maximum energies of 0.96 MeV for ^{11}C, 1.20 MeV for ^{13}N, and 1.73 MeV for ^{15}O respectively.

Solution exercise 2.11.
The main interaction process at low atomic numbers for diagnostic x-ray energies is incoherent scattering. At these energies the scattered photons have nearly the same energy as the primary ones, as little energy is transferred to the electrons. Thus the mean energy is close to 55 keV with a low attenuation in PMMA. With increasing atomic number the probability for photoelectric effect increases and dominates for atomic numbers over around 20 at the energies in the example. The photoelectric effect results in emission of characteristic x rays. For elements like iron and copper, the energies of the characteristic x rays are low, a few keV, and the photons are to a high degree absorbed in the PMMA absorber. With increasing atomic number the energy of the characteristic x rays increases and thus the transmission as well. However with very high atomic numbers, the energy of the binding energy of the K-shell electrons increases, and less photons from the 140 kV x-ray spectrum are able to emit photoelectrons from the K-shell. Instead the photoelectric effect in the L-shell increases, resulting in emission of L-x rays, with a lower energy and the transmission through the PMMA absorber decreases.

Solution 2.12.
The attenuation in the filter of iron is given by

$$\Phi(x) = \Phi(0)e^{-\mu x} \tag{2.3.19}$$

Solve the equation for x

$$x = -\frac{\ln(\Phi(x)/\Phi(0))}{\mu} \tag{2.3.20}$$

where $\Phi(x)$ is the fluence after a filter thickness of x, $\Phi(0)$ is the fluence before the filter, and μ is the linear attenuation coefficient.

There are characteristic x rays with different energies and frequencies. Below are two different approximations compared.
a) Calculation for the most common energy, $K_{\alpha 1}$(74.97 keV)

Data:

$(\mu/\rho)_{Fe}=0.06921\ m^2\ kg^{-1}$ (mass attenuation coefficient in Fe (74.97 keV))
$\rho_{Fe}=7.87\cdot10^3\ kg\ m^{-3}$ (density of Fe)
$\Phi(x)/\Phi(0)=0.1$

Data inserted in equation (2.3.20) gives

$$x = -\frac{\ln(0.1)}{0.06921 \cdot 7.87 \cdot 10^3} = 0.0042\ m$$

b)   Calculation for a weighted mean energy, $\bar{K}(76.60\ keV)$.

Data:
$(\mu/\rho)_{Fe}= 0.0658\ m^2\ kg^{-1}$ (mass attenuation coefficient in Fe (76.60 keV))

$$x = -\frac{\ln(0.1)}{0.0658 \cdot 7.87 \cdot 10^3} = 0.0044\ m$$

Answer: Thickness of iron is 4.2 mm or 4.4 mm depending on approximation.

Solution exercise 2.13.
The smallest attenuation for the secondary photons is obtained when the photons are as oblique as possible without passing more than one absorbing disk (Datched arrow in Fig. 2.19).

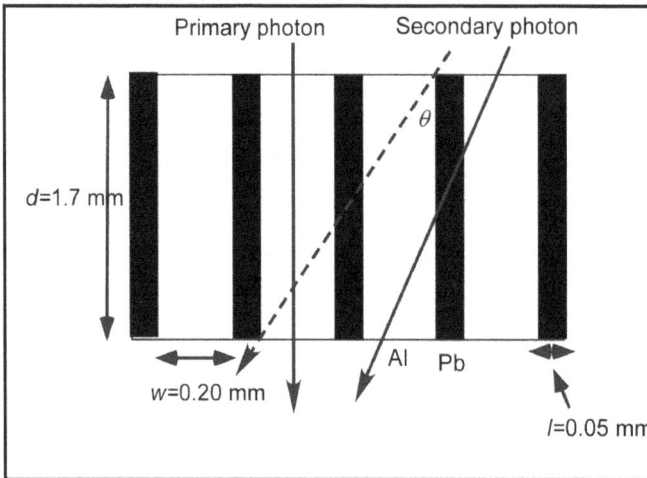

Figure 2.19: Illustration of passage of photons through a grid for diagnostic x rays.

The transmission of the primary radiation is given by

$$\Phi_p = \Phi_0 e^{-\mu_{pAl}d} \qquad (2.3.21)$$

The transmission of the secondary radiation is given by

$$\Phi_s = \Phi_0 e^{-\mu_{sAl}2w/\cos\theta} e^{-\mu_{sPb}l/\cos\theta} \qquad (2.3.22)$$

where
d=1.7 mm (thickness of the grid).
w=0.20 mm (width of the Al-strip).
l=0.05 mm (width of the Pb-strip).
Φ_0=impinging photon fluence
μ_{pAl}=0.03688 ·2.7 · 10^3 m^{-1} (linear attenuation coefficient in Al for primary radiation (50 keV))
μ_{sAl}=0.05676 ·2.7· 10^3 m^{-1} (linear attenuation coefficient in Al for secondary radiation (40 keV))
μ_{sPb}=1.431·11.34· 10^3 m^{-1} (linear attenuation coefficient in Pb for secondary radiation (40 keV))
θ=the angle between the photon and the grid. θ is obtained through the relation

$$\tan\theta = \frac{d}{2w+l} \qquad (2.3.23)$$

Data inserted gives tan θ=1.7/0.45=3.78 and θ=75.17°.

The ratio of the primary and second photon fluence is then given by

$$\frac{\Phi_p}{\Phi_s} = \frac{\Phi_0 e^{-0.03688\cdot2.7\cdot10^3\cdot0.0017}}{\Phi_0 e^{-0.05676\cdot2.70\cdot10^3\cdot2\cdot0.00020/\cos75.17} e^{-1.431\cdot11.34\cdot10^3\cdot0.00005/\cos75.17}} = 25.3$$

Answer: The ratio between primary and secondary photon fluence is 25.

Solution exercise 2.14.
The energy of the scattered photon is given by the relation

$$h\nu_s = \frac{h\nu}{1 + \frac{h\nu}{m_ec^2}(1 - \cos\theta)} \qquad (2.3.24)$$

The energy of the secondary photon shall be half of the primary photon.

Thus

$$h\nu_s = \frac{h\nu}{2} = \frac{h\nu}{1 + \frac{h\nu}{m_ec^2}(1 - \cos\theta)} \qquad (2.3.25)$$

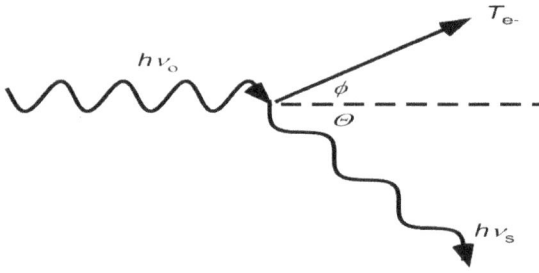

Figure 2.20: Incoherent scattering of a photon.

$$2 = 1 + \frac{hv}{m_e c^2}(1 - \cos \theta) \qquad (2.3.26)$$

$$hv = \frac{m_e c^2}{1 - \cos \theta} \qquad (2.3.27)$$

hv_{min} is obtained when $\cos\theta$=-1 and $\theta = 180°$

hv=0.511/2=0.256 MeV and θ=180°.

Answer: The electron will go in the forward direction and the photon is scattered 180°. The smallest energy for which this is possible is 0.256 MeV.

Solution exercise 2.15.
When the scattering angle is increased, the scattered photon energy is decreasing from the primary photon energy of 100 keV. With decreasing energy the mass attenuation coefficient, μ/ρ is first increased and the transmission is decreased. At 88 keV, μ/ρ suddenly drops as there is no possibility to expel K-electrons. The angle for which the scattered energy is 88 keV is given by the relation

$$hv_s = \frac{hv}{1 + \frac{hv}{m_e c^2}(1 - \cos \theta)} \qquad (2.3.28)$$

where
hv=100 keV (primary photon energy)
hv_s=88.0 keV (scattered photon energy)
$m_e c^2$=511 keV (electron rest mass energy).

Data inserted in Eq. (2.3.28) gives

$$88 = \frac{100}{1 + \frac{100}{511}(1 - \cos \theta)}$$

Solving for θ gives $\cos\theta = 0.3032$ and $\theta = 72.35°$.

The thickness of lead to obtain an increase in fluence of a factor 10 is obtained through the relation

$$f = \frac{\Phi_0 e^{-(\mu_{88-})x}}{\Phi_0^{-(\mu_{88+})x}} \tag{2.3.29}$$

where
$f = 10$ (ratio of fluences)
$\Phi_0 =$ photon fluence before Pb-filter
$\mu_{88-} = 0.1912 \cdot 11.34 \cdot 10^3$ m^{-1} (linear attenuation coefficient in Pb for an energy just below 88 keV)
$\mu_{88+} = 0.7682 \cdot 11.34 \cdot 10^3$ m^{-1} (linear attenuation coefficient in Pb for an energy just above 88 keV)
$x =$ Pb filter thickness (m).

Data inserted in Eq (2.3.29) gives

$$10 = \frac{\Phi_0 e^{-0.1912 \cdot 11.34 \cdot 10^3 x}}{\Phi_0^{-0.7682 \cdot 11.34 \cdot 10^3 x}}$$

$x = 0.35$ mm.

Answer: The increase in counting rate is obtained at 72°. The thickness of the Pb-filter is 0.35 mm.

Solution exercise 2.16.
The relation between the Compton scattered photon and the primary photon is given by

$$h\nu_s = \frac{h\nu}{1 + \frac{h\nu}{m_e c^2}(1 - \cos\theta)} \tag{2.3.30}$$

where $h\nu$ is the primary photon energy, $h\nu_s$ is the scattered photon energy and $m_e c^2$ is the electron rest mass energy (0.511 MeV).

Equation (2.3.30) can be rearranged by inverting, to give

$$\frac{1}{h\nu_s} = \frac{1}{h\nu} + \frac{1}{m_e c^2}(1 - \cos\theta) \tag{2.3.31}$$

The last factor is independent of the photon energy. This implies that in general the scattered photon energy after n equal scatterings in the angle θ/n will be given by the relation

$$\frac{1}{h\nu_n} = \frac{1}{h\nu} + \frac{n}{m_e c^2}(1 - \cos(\theta/n)) \tag{2.3.32}$$

In the example the total scattering angle θ is π radians obtained in 1, 2, 3 or infinite number of equal scatterings for a primary photon energy of 1.17 MeV. Data inserted gives

$n=1$

$$\frac{1}{hv_1} = \frac{1}{1.17} + \frac{1}{0.511}(1 - \cos(\pi))$$

$$\frac{1}{hv_1} = \frac{1}{1.17} + \frac{1}{0.511}(1 + 1)$$

$hv_1 = 0.210$ MeV

$n=2$

$$\frac{1}{hv_2} = \frac{1}{1.17} + \frac{2}{0.511}(1 - \cos(\pi/2))$$

$$\frac{1}{hv_2} = \frac{1}{1.17} + \frac{2}{0.511}(1 - 0)$$

$hv_2 = 0.210$ MeV

Note that the photon energy after 1 or 2 scatterings here is exactly the same.

$n=3$

$$\frac{1}{hv_3} = \frac{1}{1.17} + \frac{3}{0.511}(1 - \cos(\pi/3))$$

$$\frac{1}{hv_3} = \frac{1}{1.17} + \frac{3}{0.511}(1 - 0.5)$$

$hv_3 = 0.264$ MeV

When n becomes large, π/n will become small and $\cos(\pi/n)$ may be written as a series

$$\cos(\pi/n) = 1 - \frac{\pi^2}{n^2 2!} + \frac{\pi^4}{n^4 4!} - \dots \qquad (2.3.33)$$

This gives

$$\frac{1}{hv_n} = \frac{1}{hv} + \frac{n}{m_e c^2}\left(1 - 1 + \frac{\pi^2}{n^2 2!} - \frac{\pi^4}{n^4 4!} + \dots\right) \qquad (2.3.34)$$

When $n \to \infty$ then the last factor in the equation goes to zero and $1/hv_n = 1/hv$.

Thus

$$\lim_{n \to \infty} \frac{1}{hv_n} = \frac{1}{hv} \qquad (2.3.35)$$

$hv_\infty = 1.17$ MeV

Answer: The energy of the scattered photon after being scattered in 180° is a) 0.21 MeV(n=1), b) 0.21 MeV(n=2), c) 0.26 Mev (n=3) and d) 1.17 MeV ($n = \infty$).

Solution exercise 2.17.

a) The kinetic energies of the photoelectrons are given by

$$T' = h\nu - B_{K,L,M} \tag{2.3.36}$$

where $B_{K,L,M}$ are the the binding energies of the K, L and M electrons respectively. This energy is independent of the emission angle. As example B_K=0.088 MeV. This gives

$$T_K=2.04\text{-}0.088=1.95\,\text{MeV}$$

b) The relation between the angle of the scattered photon, θ, and the angle, ϕ, of the emitted Compton electron is e.g. given in Physics Handbook (Nordling and Osterman, 2006)

$$\cos\theta = 1 - \frac{2}{(1 + h\nu/m_e c^2)^2 \tan^2\phi + 1} \tag{2.3.37}$$

where
$h\nu$=2.04 MeV (photon energy)
$m_e c^2$=0.511 MeV (electron rest mass)
ϕ=20° (emission angle of the Compton electron)

Data inserted in Eq. (2.3.37) gives

$$\cos\theta = 1 - \frac{2}{(1 + 2.04/0.511)^2 \tan^2 20 + 1}$$

$\cos\theta$=0.536

The energy of the compton electron is given by

$$T' = \frac{h\nu(h\nu/m_e c^2)(1 - \cos\theta)}{1 + (h\nu/m_e c^2)(1 - \cos\theta)} \tag{2.3.38}$$

Data inserted Eq. (2.3.38) gives

$$T' = \frac{2.04 \cdot 2.04/0.511(1 - 0.535)}{1 + 2.04/0.511(1 - 0.535)} = 1.33\,\text{MeV}$$

c) In pair production the photon energy is shared between the electron and the positron in all configurations. However, an energy corresponding to $2m_e c^2$ is used to produce the electron-positron pair and is not available as kinetic energy. Thus the kinetic energy to be shared between the electron and the positron is

$$T' = h\nu - 2m_e c^2 = 2.04 - 2 \cdot 0.511 = 1.02\,\text{MeV} \tag{2.3.39}$$

This holds if the influence of the charge of the nucleus is neglected. In general the energy distribution between the electron and the positron has a nearly uniform shape with the same probabilities for the two particles. However, in particular at low energies, the positive charge will have an attractive force on the electron and a repulsive force on the positron. All possible energies, between zero and 1.02 MeV can be obtained in all emission angles. In the emission angle of 20° there is a slight larger probability for the higher energies.

Answer: The electron energies are a) 1.91 MeV and higher depending on the interactive electron shell, b) 1.33 MeV, c) 0-1.02 MeV.

Solution exercise 2.18.

μ_{tr}/ρ is defined as (ICRU, 2011)

$$\frac{\mu_{tr}}{\rho} = \frac{\mu}{\rho}f$$

where

$$f = \frac{\sum f_J \sigma_J}{\sum \sigma_J}$$

and f_J is the average fraction of the incident particle energy that is transferred to kinetic energy of charged particles in an interaction type J.

For photons this can be written as

$$\mu_{tr}/\rho - (\mu/\rho)(\bar{T}/h\nu) \tag{2.3.40}$$

where $\bar{T}$ is the mean energy of the kinetic energies of the emitted electrons. To calculate the mean energy of the emitted electrons it is necessary to make separate calculations for all different interaction probabilities; photoelectric effect, incoherent scattering and, pair production in the nuclear and the electron field.

a) Photoelectric effect.

The electrons emitted in the photoelectric effect are photoelectrons and Auger electrons including Coster-Cronig electrons. The energy not obtained as kinetic energy of the electrons is obtained as energy of fluorescence x rays. It is more practical to calculate the photon energy emitted and deduct it from the primary photon energy, to get the electron energy. Then the mass energy transfer coefficient τ_{tr}/ρ for the photoelectric effect can be written as

$$\tau_{tr}/\rho = (\tau/\rho)\left[1 - \frac{p_K \omega_K h\nu_K}{h\nu} - \frac{(1-p_K)p_L \omega_L h\nu_L}{h\nu} - \cdots\right] \tag{2.3.41}$$

where

$\tau/\rho = 0.503 \cdot 10^{-3}$ m^2 kg^{-1} (mass cross section for photoelectric effect (2.0 Mev))

p_K=0.79 (probability for a photoelectric effect in the K-shell)

p_L=0.75 (probability for a photoelectric effect in the L_I-shell, for the photons not interacting in the K-shell. There are three L-subshells, but in these calculations only the L_I-shell is included. As the contribution from the L-shells is small, this approximation will not affect the numerical value significantly)

ω_K=0.96 (fluorescence yield for K-shell)

ω_L=0.32 (fluorescence yield for the L_I-shell)

$h\nu_K$=75 keV (mean energy of the K-x rays)

$h\nu_L$=12 keV (mean energy of the L-x rays)

$h\nu$=2.0 Mev (primary photon energy)

Data inserted in Eq. (2.3.41) gives

$$\tau_{tr}/\rho = (\tau/\rho) \left[1 - \frac{0.79 \cdot 0.96 \cdot 0.075}{2.0} - \frac{(1 - 0.79)0.75 \cdot 0.32 \cdot 0.012}{2.0} - \cdots \right]$$

$\tau_{tr}/\rho = 0.503 \cdot 10^{-3} \cdot 0.969 = 0.487 \cdot 10^{-3}\ \text{m}^2\ \text{kg}^{-1}$

b) Incoherent scattering

To calculate the mean energy of the electrons emitted in incoherent scattering, it is possible in the first approximation, to use the Klein-Nishina relation to obtain the electron energy distribution. This mean energy is then independent of the atomic number. The value is taken from Table 2.1.

The mean electron energy obtained for 2 MeV photons is $\bar{T}_{incoh}$=1.062 MeV

$(\sigma/\rho) = 0.348 \cdot 10^{-2}\ \text{m}^2\ \text{kg}^{-1}$ (Klein-Nishina cross section (2 MeV))

Data inserted in Eq. (2.3.40) for incoherent scattering gives

$\sigma_{tr}/\rho = 0.348 \cdot 10^{-2}(1.062/2.0) = 0.185 \cdot 10^{-2}\ \text{m}^2\ \text{kg}^{-1}$

c) Pair production

In pair production the electron and positron will obtain the photon energy, except the energy needed to produce the positron and the electron, $2m_ec^2$. Thus

$$\bar{T}_{pair} = h\nu - 2m_ec^2 \tag{2.3.42}$$

$(\kappa/\rho) = 0.545 \cdot 10^{-3}\ \text{m}^2\ \text{kg}^{-1}$ (mass cross section for pair production (2 MeV))

Data inserted in Eq. (2.3.40) for pair production gives

$$\kappa_{tr}/\rho = 0.545 \cdot 10^{-3} \frac{2.0-1.022}{2.0} = 0.267 \cdot 10^{-3} \, m^2 \, kg^{-1}$$

The total mass energy transfer coefficient is now obtained by adding all separate contributions

$$\mu_{tr}/\rho = \tau_{tr}/\rho + \sigma_{tr}/\rho + \kappa_{tr}/\rho \tag{2.3.43}$$

Thus

$$\mu_{tr}/\rho = 0.487 \cdot 10^{-3} + 1.85 \cdot 10^{-3} + 0.267 \cdot 10^{-3} = 2.60 \cdot 10^{-3} \, m^2 \, kg^{-1}$$

Corresponding data in Higgins et al (Higgins et al, 1991) are

$$\mu_{tr}/\rho = 0.49 \cdot 10^{-3} + 1.81 \cdot 10^{-3} + 0.27 \cdot 10^{-3} = 2.57 \cdot 10^{-3} \, m^2 \, kg^{-1}.$$

There is a good agreement between the data in these calculations and the more extensive ones in the table by Higgins in spite of the approximations. The main difference is for incoherent scattering probably due to the assumption that the electron is free and at rest in our calculation, while the table by Higgins includes the correction for electron binding energies.

Answer: The mass energy transfer coefficient for 2.0 Mev photons in Pb is $2.57 \cdot 10^{-3} \, m^2 \, kg^{-1}$.

Solution exercise 2.19.

Primary photon fluence at (A)

Assume that the photon fluence just above the beam flattening filter is Φ_0. Attenuation and the inverse square law will give that the primary photon fluence at (A) is

$$\Phi_p = \frac{\Phi_0 l_1^2 e^{-\mu_p d}}{(l_1 + l_2)^2} \tag{2.3.44}$$

where l_1 is the distance from radiation source to filter, l_2 is the distance from filter to phantom surface, μ_p is the linear attenuation coefficient in Pb and d is the filter thickness.

Secondary photon fluence at (B)

Assume that all secondary photons are produced at the central axis. This is a

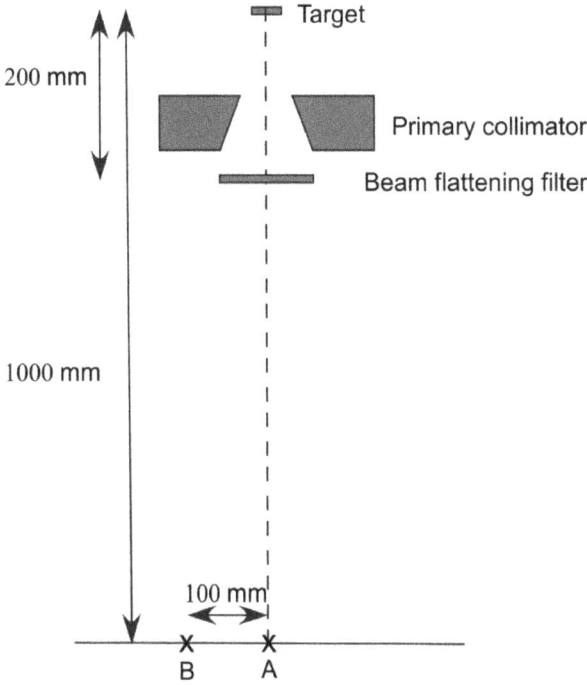

Figure 2.21: Illustration of the treatment head used in the calculations in exercise 2.19.

very good approximation considering the geometry. All coherent scattered photons are considered as primary photons as they have the same energy as the primary photons and are emitted at rather small angles. There are thus three possible interaction processes, photoelectric effect, incoherent scattering and pair production. The contribution from photoelectric effect is neglected as the attenuation of the produced fluorescence x rays is high.

a) Incoherent scattering

The fluence of scattered photons at (B) is obtained by integrating over the total thickness, taking into consideration the attenuation of both the primary and the secondary radiation.

$$\Phi_{\text{s,incoh}} = \frac{\int_0^d \Phi_0 (\frac{d_e\sigma}{d\Omega})_\theta S_1 \rho_{\text{Pb}} N_e e^{-\mu_p x} e^{-\mu_{s1}(d-x)/\cos\theta}\, dx}{(l_2/\cos\theta)^2} \tag{2.3.45}$$

Solving the integral gives

$$\Phi_{\text{s,incoh}} = \frac{\Phi_0 (\frac{d_e\sigma}{d\Omega})_\theta S_1 N_e e^{-\mu_{s1}(d/\cos\theta)}(1 - e^{-d(\mu_p - \mu_{s1}/\cos\theta)})}{(l_2/\cos\theta)^2((\mu_p/\rho) - (\mu_{s1}/\rho)/\cos\theta)} \tag{2.3.46}$$

b) Pair production

The fluence of annihilation photons at (B), assuming that all produced positrons are annihilated in the filter, is obtained in a corresponding way.

$$\Phi_{s,pair} = \frac{\int_0^d \Phi_0 2(\kappa/\rho)\rho_{Pb}s_1 e^{-\mu_p x}e^{-\mu_{s2}(d-x)/\cos\Theta}\,dx}{4\pi(l_2/\cos\theta)^2} \qquad (2.3.47)$$

Solving the integral gives

$$\Phi_{s,pair} = \frac{\Phi_0 2(\kappa/\rho)s_1 e^{-\mu_{s2}d/\cos\Theta}(1 - e^{-d(\mu_p-\mu_{s2}/\cos\Theta)})}{4\pi(l_2/\cos\theta)^2((\mu_p/\rho) - (\mu_{s2}/\rho)/\cos\Theta)} \qquad (2.3.48)$$

Data:
l_1=0.200 m (distance from radiation source to filter)
l_2=0.800 m (distance from filter to phantom surface)
d=0.007 m (filter thickness)
$\frac{d_e\sigma}{d\Omega}$=74.0·$10^{-31}$ m²/steradian per electron (Klein-Nishina cross section for scattering in angle θ (2.0 Mev))
θ=7° (angle at which the photons are emitted to reach point (B))
s_1=0.025²·π m² (area of the beam flattening filter)
N_e=2.384·10^{26} (number of electrons per kg in Pb)
$h\nu_{s1}$=1.94 MeV (incoherent scattered photons)
$h\nu_{s2}$=0.511 MeV (annihilation photons)
μ_p/ρ=4.60·10^{-3} m² kg^{-1} (mass attenuation coefficient for the primary photons (2.0 Mev) in Pb)
μ_{s1}/ρ=4.65·10^{-3} m² kg^{-1} (mass attenuation coefficient for the scattered photons (1.94 MeV) from the incoherent scattering in Pb)
μ_{s2}/ρ=1.572·10^{-2} m² kg^{-1} (mass attenuation coefficient for the annihilation photons (0.511 MeV) from the pair production in Pb)
κ/ρ=5.45·10^{-4} m² kg^{-1} (mass cross section for pair production in Pb (2 MeV))
ρ_{Pb}=11.34·10^3 kg m^{-3} (density of Pb)

Data inserted in Eq. (2.3.44), (2.3.46) and (2.3.48) gives

a) Primary photon fluence

$$\Phi_p = \frac{\Phi_0 \cdot 0.200^2 e^{-4.60\cdot10^{-3}\cdot0.007\cdot11.34\cdot10^3}}{1.0^2} = 2.776 \cdot 10^{-2}\Phi_0 \text{ m}^{-2}$$

b) Incoherent scattering

$$\Phi_{s,incoh} = \frac{\Phi_0 \cdot 0.025^2 \cdot \pi \cdot 74.0 \cdot 10^{-31} \cdot 2.384 \cdot 10^{26} \cdot e^{-4.65\cdot10^{-3}\cdot11.34\cdot10^3(0.007/\cos 7)}}{(0.8/\cos 7)^2(4.60 \cdot 10^{-3} - 4.65 \cdot 10^{-3}/\cos 7)}$$

$$\times (1 - e^{-0.007 \cdot 11.34 \cdot 10^3 (4.60 \cdot 10^{-3} - 4.65 \cdot 10^{-3}/\cos 7)})$$

$$\Phi_{s,incoh} = 2.928 \cdot 10^{-4} \Phi_0 \text{ m}^{-2}$$

c) Pair production

$$\Phi_{s,pair} = \frac{\Phi_0 2 \cdot 5.45 \cdot 10^{-4} \cdot 0.025^2 \cdot \pi \cdot e^{-1.572 \cdot 10^{-2} \cdot 11.34 \cdot 10^3 (0.007/\cos 7)}}{4\pi(0.8/\cos 7)^2 (4.60 \cdot 10^{-3} - 1.572 \cdot 10^{-2}/\cos 7)}$$
$$\times (1 - e^{-0.007 \cdot 11.34 \cdot 10^3 (4.60 \cdot 10^{-3} - 1.572 \cdot 10^{-2}/\cos 7)})$$

$$\Phi_{s,pair} = 9.56 \cdot 10^{-6} \Phi_0 \text{ m}^{-2}$$

$$\Phi_s = \Phi_{s,incoh} + \Phi_{s,pair} = 2.928 \cdot 10^{-4} + 9.56 \cdot 10^{-6} = 3.023 \cdot 10^{-4} \Phi_0 \text{ m}^{-2}$$

The ratio between the secondary and primary photon fluence is thus

$$\frac{\Phi_s}{\Phi_p} = \frac{3.023 \cdot 10^{-4} \Phi_0}{2.776 \cdot 10^{-2} \Phi_0} = 1.09 \cdot 10^{-2}$$

Answer: The ratio of the secondary and the primary photon fluence is 0.011.

Solution exercise 2.20.
The electron fluence just behind the disk from a thin slice dx is given by

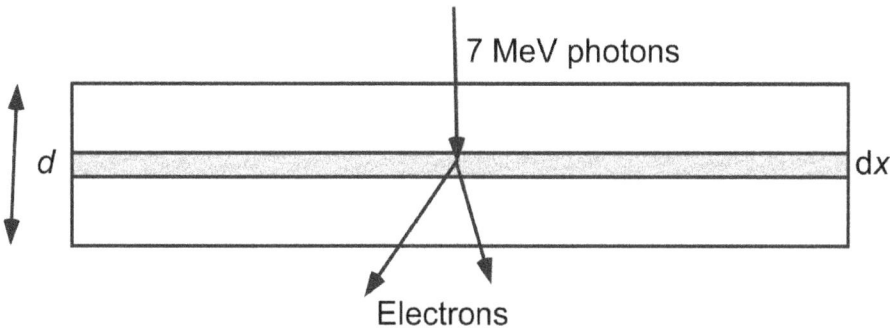

Figure 2.22: Sketch of the PMMA sheet in exercise 2.20 where the electrons are produced.

$$d\Phi_e = \Phi(\sigma/\rho)e^{-(\mu/\rho)\rho x}e^{-(\beta_1/\rho)\rho(d-x)}\rho \, dx + \Phi 2(\kappa/\rho)e^{-(\mu/\rho)\rho x}e^{-(\beta_2/\rho)\rho(d-x)}\rho \, dx \qquad (2.3.49)$$

The first factor describes the contribution from incoherent scatter (one electron) and the second factor the contribution from pair production (one electron and one positron. The third electron in the pair production in the electron field is neglected as it often gets a very low energy and thus will be absorbed in the disk. The contribution from photoelectric effect can be neglected for this high photon energy.

The total electron fluence is obtained by integrating over the whole disk. Integration over the thickness d gives

$$\Phi_e = \Phi(\sigma/\rho)\frac{e^{-(\mu/\rho)\rho d} - e^{-(\beta_1/\rho)\rho d}}{\beta_1/\rho - \mu/\rho} + \Phi 2(\kappa/\rho)\frac{e^{-(\mu/\rho)\rho d} - e^{-(\beta_2/\rho)\rho d}}{\beta_2/\rho - \mu/\rho} \qquad (2.3.50)$$

where Φ is the photon fluence just above the disk, σ/ρ is the mass incoherent scattering cross section, κ/ρ is the mass pair production cross section, μ/ρ is the mass attenuation coefficient, β_1/ρ is the mass absorption coefficient for the electrons from the incoherent scatter, and β_2/ρ is the mass absorption coefficient for the electrons from the pair production.

The same absorption coefficient for electrons and positrons is assumed

a) Disk of PMMA

Data:
$(\sigma/\rho)_{PMMA} = 2.144 \cdot 10^{-3}$ m^2 kg^{-1} (mass scattering cross section for 7 MeV)
$(\kappa/\rho)_{PMMA} = 3.22 \cdot 10^{-4}$ m^2 kg^{-1} (mass pair production cross section for 7 MeV)
$(\mu/\rho)_{PMMA} = 2.47 \cdot 10^{-3}$ m^2 kg^{-1} (mass attenuation coefficient for 7 MeV)
$\rho d = 15$ kg m^{-2} (thickness of PMMA)
$\Phi = 1.0 \cdot 10^{15}$ m^{-2} (photon fluence at disk)

The mass absorption coefficient for the electrons is obtained from the relation

$$\beta_{PMMA}/\rho = \frac{3.5Z}{AE^{1.14}} \text{ m}^2\text{kg}^{-1}$$

PMMA has the chemical composition ($C_5H_8O_2$). This gives $Z/A = 54/100 = 0.54$.

For incoherent scatter the maximum electron energy, that shall be used in the equation is obtained, when the scattered photon has its lowest energy, i.e. when the electron is emitted in the forward direction and the photon is scattered in $180°$. The electron energy is thus with $h\nu = 7.0$ Mev

$$T = 7.0 - \frac{7.0}{1 + \frac{7.0}{0.511}(1 - \cos 180)}$$

This gives

T=6.75 MeV

and

$$\beta_{PMMA,1}/\rho = \frac{3.5 \cdot 0.54}{6.75^{1.14}} = 0.214 \text{ m}^2\text{kg}^{-1}$$

For pair production the electron energy is obtained by assuming that the total available energy is given to either the electron or the positron. The available energy is equal to the photon energy reduced with $2m_ec^2$, 1.022 MeV. Thus

T=7.0-1.022=5.98 MeV

$$\beta_{PMMA,2}/\rho = \frac{3.5 \cdot 0.54}{5.98^{1.14}} = 0.246 \text{ m}^2\text{kg}^{-1} \qquad (2.3.51)$$

Data inserted in Eq. (2.3.50) gives

$$\Phi_e = 1.0 \cdot 10^{15} \cdot 2.144 \cdot 10^{-3} \frac{e^{-2.47 \cdot 10^{-3} \cdot 15} - e^{-0.214 \cdot 15}}{0.214 - 2.47 \cdot 10^{-3}}$$
$$+ 1.0 \cdot 10^{15} \cdot 2 \cdot 3.22 \cdot 10^{-4} \frac{e^{-2.47 \cdot 10^{-3} \cdot 15} - e^{-0.246 \cdot 15}}{0.246 - 2.47 \cdot 10^{-3}} \text{ m}^{-2}$$

The electron fluence with a PMMA disk is Φ_e=1.184·10^{13} m^{-2}.

b) Disk of lead

Data:
$(\sigma/\rho)_{Pb}$=1.57·10^{-3} m^2 kg^{-1} (mass scattering cross section for 7 MeV)
$(\kappa/\rho)_{Pb}$=2.87·10^{-3} m^2 kg^{-1} (mass pair production cross section for 7 MeV)
$(\mu/\rho)_{Pb}$=4.53·10^{-3} m^2 kg^{-1} (mass attenuation coefficient for 7 MeV)

The mass absorption coefficient for the electrons is obtained from the equation

$$\beta_{Pb}/\rho = \frac{0.77Z^{0.31}}{E^{1.14}}$$

Here Z=82 and the same electron energies as for the calculations for PMMA may be used.

Incoherent electrons:

$$\beta_{Pb,1}/\rho = \frac{0.77 \cdot 82^{0.31}}{6.75^{1.14}} = 0.342 \text{ m}^2\text{kg}^{-1}$$

Pair production electrons:

$$\beta_{Pb,2}/\rho = \frac{0.77 \cdot 82^{0.31}}{5.98^{1.14}} = 0.393 \text{ m}^2\text{kg}^{-1}$$

These data inserted in the Eq. (2.3.50) gives

$$\Phi_e = 1.0 \cdot 10^{15} \cdot 1.57 \cdot 10^{-3} \frac{e^{-4.53 \cdot 10^{-3} \cdot 15} - e^{-0.342 \cdot 15}}{0.342 - 4.53 \cdot 10^{-3}}$$

$$+ 1.0 \cdot 10^{15} \cdot 2 \cdot 2.87 \cdot 10^{-3} \frac{e^{-4.53 \cdot 10^{-3} \cdot 15} - e^{-0.393 \cdot 15}}{0.393 - 4.53 \cdot 10^{-3}} \ \text{m}^{-2}$$

The electron fluence with a lead disk is $\Phi_e = 1.81 \cdot 10^{13}$ m^{-2}.

Answer: The electron fluence is for a PMMA disk $1.18 \cdot 10^{13}$ m^{-2} and for a lead disk $1.81 \cdot 10^{13}$ m^{-2}.

Solution exercise 2.21.
 a) Sphere of lead The dominating interaction process for a photon with an energy

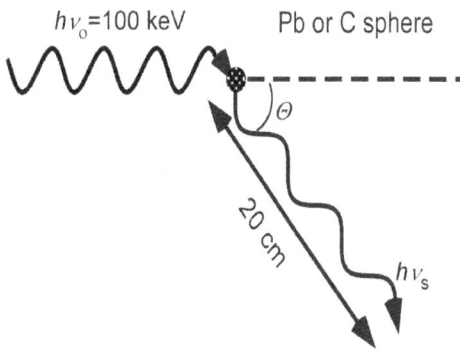

Figure 2.23: Irradiation geometry in exercise 2.21.

of 100 keV in lead is the photoelectric effect. The fluence rate of secondary K-x rays, neglecting self-absorption in the sphere, is given by the relation

$$\dot{\Phi}_{s,Pb} = \frac{\dot{\Phi}_0 m(\tau_K/\rho)\omega_K}{4\pi l^2} \tag{2.3.52}$$

where $\dot{\Phi}_{s,Pb}$ is the fluence rate of secondary K x rays at the measuring point, $\dot{\Phi}_0$ is the fluence rate of primary photons at the center of the sphere, m is the mass of the sphere, τ_K/ρ is the mass cross section for photoelectric effect in the K-shell, ω_K is the fluorescence yield of K-x rays and l is the distance from sphere to measuring point.

b) Sphere of carbon
Only incoherent scattered photons are included in the calculations. The fluence rate

of these photons at angle θ, neglecting self-absorption in the sphere is given by the relation

$$\dot{\Phi}_{s,Cs} = \frac{\dot{\Phi}_o m N_e (\frac{d_e\sigma}{d\Omega})_\theta}{l^2} \qquad (2.3.53)$$

where N_e is $N_A Z/m_a$ (number of electrons per mass unit) and $(d_e\sigma/d\Omega)_\theta$ is the Klein-Nishina cross section per electron. The ratio between the fluence with a carbon sphere and a lead sphere is then

$$\frac{\dot{\Phi}_{s,Cs}}{\dot{\Phi}_{s,Pb}} = \frac{\dot{\Phi}_o m N_e (\frac{d_e\sigma}{d\Omega})_\theta 4\pi l^2}{\dot{\Phi}_o m (\tau_K/\rho)\omega_K l^2} \qquad (2.3.54)$$

After simplification

$$\frac{\dot{\Phi}_{s,Cs}}{\dot{\Phi}_{s,Pb}} = \frac{N_e (\frac{d_e\sigma}{d\Omega})_\theta 4\pi}{(\tau_K/\rho)\omega_K} \qquad (2.3.55)$$

The angle θ should be chosen in order to have the same energy of the incoherent scattered photons and the $K_{\alpha II}$ x rays.

The energy of the scattered photons is obtained from the Compton equation.

$$h\nu_s = \frac{h\nu}{1 + \frac{h\nu}{m_e c^2}(1 - \cos\theta)} \qquad (2.3.56)$$

The energy of $K_{\alpha II}$ x rays for lead is 74.97 keV. This is inserted in Eq. (2.3.56)

$$74.97 = \frac{100}{1 + \frac{100}{511}(1 - \cos\theta)} \qquad (2.3.57)$$

This gives
$\cos\theta = -0.706$ and $\theta = 135°$

Data:
$(d_e\sigma/d\Omega)_{135} = 35.3\cdot10^{-31}$ m^2/steradian (Klein-Nishina cross section for 100 keV)
$\omega_K = 0.97$ (fluorescence yield in Pb for 100 keV)
$\tau_K/\rho = (\tau_K/\tau)(\tau/\rho) = 0.788\cdot0.524$ m^2 kg^{-1} (mass photoelectric cross section in C for 100 keV)
$N_e = 6.022\cdot10^{26} \cdot 6/12$ (number of electrons per kg in C)

Data inserted in Eq. (2.3.55) gives

$$\frac{\dot{\Phi}_{s,C}}{\dot{\Phi}_{s,Pb}} = \frac{6.022\cdot10^{26}\cdot6\cdot35.3\cdot10^{-31}4\pi}{12\cdot0.788\cdot0.524\cdot0.97}$$

$\dot{\Phi}_{s,C}/\dot{\Phi}_{s,Pb} = 3.33\cdot10^{-2}$

Answer: The ratio between the fluence rates with a sphere of carbon and a sphere of lead is 0.033.

Solution exercise 2.22.
The total filter thickness is obtained from the equations below

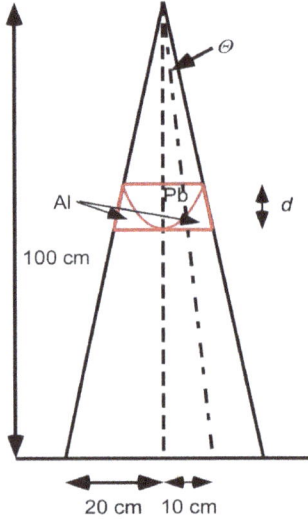

Figure 2.24: Irradiation geometry with the beam flattening filter in exercise 2.22.

At central axis:
$$\Phi_c = \Phi(0)e^{-d\mu_{Pb,7}} \tag{2.3.58}$$

At field edge:
$$\Phi_e = \Phi(\theta)e^{-d\mu_{Al,6}/\cos\theta} \tag{2.3.59}$$

$\Phi_c = \Phi_e$ (for full flattening)

Thus
$$\Phi(0)e^{-d\mu_{Pb,7}} = \Phi(\theta)e^{-d\mu_{Al,6}/\cos\theta} \tag{2.3.60}$$

and
$$d = \frac{\ln(\Phi(\theta)/\Phi(0))}{\mu_{Al,6}/\cos\theta - \mu_{Pb,7}} \tag{2.3.61}$$

where $\Phi(0)$ is the fluence at central axis before the filter, $\Phi(\theta)$ is the fluence at field edge before the filter, d is the filter thickness and θ is the emission angle at field edge. $\mu_{Pb,7}=0.00453\cdot11.34\cdot10^3$ m^{-1} (linear attenuation coefficient in Pb for 7 Mev)

$\mu_{Al,6}=0.00266{\cdot}2.70\cdot 10^{3}$ m^{-1} (linear attenuation coefficient in Al for 6 Mev)

The fluence at the edge of the field is given by the equation

$$\Phi(\theta) = \frac{\Phi(0)}{1 + \frac{(T\theta)^{1.4}}{1.73^{1.4}}} \tag{2.3.62}$$

$T = 21$ MeV (electron energy in MeV)

The angle θ is obtained from $\tan \theta = 20/100$ and $\theta=0.197$ rad$=11.29°$

Thus

$$\Phi(\theta) = \frac{\Phi(0)}{1 + \frac{(21{\cdot}0.197)^{1.4}}{1.73^{1.4}}} = 0.228\Phi(0) \tag{2.3.63}$$

Data inserted in Eq. (2.3.61) gives

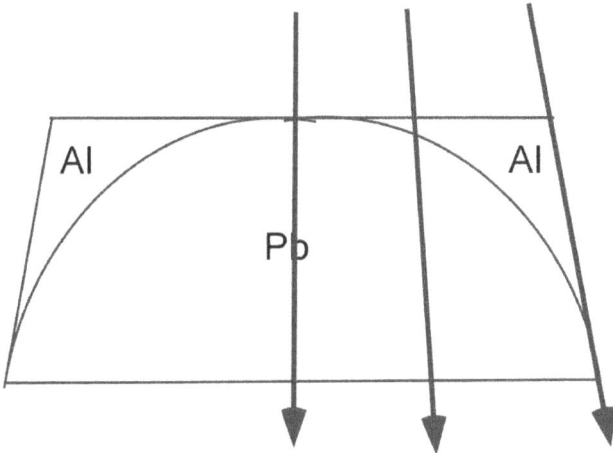

Figure 2.25: Transmission through the beam flattening filter.

$$d = \frac{\ln 0.228}{0.00266 \cdot 2.70 \cdot 10^{3}/\cos 11.29 - 0.00453 \cdot 11.34 \cdot 10^{3}} = 33.56 \text{ mm} \tag{2.3.64}$$

d=33.56 mm

To obtain the same fluence at a distance of 10 cm from the central axis the following relation must hold (assuming that the photon energy decreases linearly with distance from the central axis).

$$\Phi(\alpha)e^{-t\mu_{Al,6.5}/\cos \alpha}e^{-(d-t)\mu_{Pb,6.5}/\cos \alpha} = \Phi(0)e^{-\mu_{Pb,7}d} \tag{2.3.65}$$

Solving t gives

$$t = \frac{\ln(\Phi(\alpha)/\Phi(0))e^{d(\mu_{Pb,6.5}/\cos_\alpha - \mu_{Pb,7})}}{(\mu_{Al,6.5} - \mu_{Pb,6.5}/\cos \alpha)} \tag{2.3.66}$$

where $\Phi(\alpha)$ is the fluence half way to the edge of the field before the filter, α is the opening angle at 10 cm from the central axis and t is the thickness of Al-layer half way to the edge of the field.

The angle α is obtained from $\tan \alpha = 10/100$ and α=0.10 rad=5.73°

The fluence, $\Phi(\alpha)$, at the 10 cm from the central axis is then given by the equation

$$\Phi(\alpha) = \frac{\Phi(0)}{1 + \frac{(21 \cdot 0.10)^{1.4}}{1.73^{1.4}}} \tag{2.3.67}$$

$\Phi(0.10)=0.4326\Phi(0)$

$\mu_{Pb,6.5}$=0.00446·11.34 · 10^3 m^{-1} (linear attenuation in Pb for 6.5 Mev)
$\mu_{Al,6.5}$=0.00259·2.70 · 10^3 m^{-1} (linear attenuation in Al for 6.5 Mev)

Data inserted in Eq. (2.3.66) gives

$$t = \frac{\ln(0.4326)e^{0.03356 \cdot 11.34 \cdot 10^3 (0.00446/\cos 0.10 - 0.00453)}}{(0.00259 \cdot 2.70 \cdot 10^3 - 0.00446 \cdot 11.34 \cdot 10^3/\cos 5.73)} = 0.0188 \, m \tag{2.3.68}$$

This implies that the thickness of Pb= 33.5-18.8=14.7 mm.

The ratio of the thicknesses of Pb and Al: 14.7/19.5=0.75

Answer: The ratio between the lead and aluminum thicknesses is 0.75 .

Solution exercise 2.23.
The energies after the second disk shall be 80±10 keV or 400±50 keV. Which type of interactions in the disks will result in these energies?

a) Incoherent scattering in the first disk and the second disk.

The energy of the scattered photons is given by the Compton equation.

$$h\nu_s = \frac{h\nu}{1 + \frac{h\nu}{m_e c^2}(1 - \cos \theta)} \tag{2.3.69}$$

With hν=50 MeV and θ=45° the energy of the scattered photons is

$$h\nu_{s1} = \frac{50}{1 + \frac{50}{0.511}(1 - \cos 45)} = 1.686 \, \text{MeV}$$

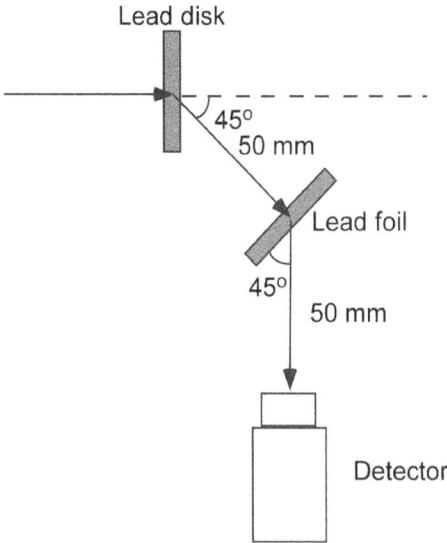

Figure 2.26: Irradiation geometry in exercise 2.23

Assume that these photons are scattered once more in $45°$ in the second disk. The energy will then become

$$hv_{s2} = \frac{1.686}{1 + \frac{1.686}{0.511}(1 - \cos 45)} = 0.86\,\text{MeV}$$

This is not an energy within the given ranges and will not be considered.

b) Pair production in the first disk and incoherent scattering in the second disk.

The energy of the secondary photons from the first disk is $0.511\,\text{MeV}$ (annihilation photons). Incoherent scattering of these photons in the second disk will give

$$hv_{s3} = \frac{0.511}{1 + \frac{0.511}{0.511}(1 - \cos 45)}$$

$hv_{s3} = 0.395\,\text{MeV}$ (within the range)

c) Photoelectric effect in the second disk.

The energy of the fluorescence x rays is around $80\,\text{keV}$ and within the given range.

Calculation of the fluence rate.

Contribution of photons in the 400±50 keV range is thus obtained from a pair production in the first disk and incoherent scattering in the second disk.

The fluence rate of annihilation photons at the second disk is given by (assuming all positrons are annihilated)

$$\dot{\Phi}_{sp} = \dot{\Phi}_0 \rho d_1 S_1 \frac{2\kappa/\rho}{4\pi l_1^2} \tag{2.3.70}$$

where $\dot{\Phi}_0$ is the impinging fluence rate at the first disk, d_1 is the thickness of the first disk, S_1 is the beam area, l_1 is the distance from the first disk to the second disk, κ/ρ is the mass cross section for pair production and ρ is the density of lead.

The fluence rate of scattered photons in the energy range 400 ± 50 keV at the detector is then given by

$$\dot{\Phi}_s = \dot{\Phi}_{sp} N_0 d_2 S_2 \frac{d_e\sigma}{d\Omega}\frac{1}{l_2^2} = \dot{\Phi}_0 \rho d_1 S_1 \frac{2\kappa/\rho}{4\pi l_1^2} N_0 d_2 S_2 \frac{d_e\sigma}{d\Omega}\frac{1}{l_2^2} \tag{2.3.71}$$

where $\dot{\Phi}_{sp}$ is the impinging fluence rate of annihilation photons at the second disk, N_0 is the number of electrons/volume unit in the disk, d_2 is the thickness of the second disk, $d_e\sigma/d\Omega$ is the Klein-Nishina cross section, l_2 is the distance from the second disk to the detector and S_2 is the disk area.

Data:
$\dot{\Phi}_0 = 3.0 \cdot 10^{10}$ m^{-2}s^{-1}
$d_1 = 10$ mm
$S_1 = \pi \cdot 3^2$ mm^2
$l_1 = 50$ mm
$\kappa/\rho = 0.770 \cdot 10^{-2}$ m^2 kg^{-1} (50 MeV)
$N_0 = 6.022 \cdot 10^{26} \cdot 11.34 \cdot 10^3 \cdot 82/208$ (number of electrons per m^3)
$d_2 = 0.10$ mm
$d_e\sigma/d\Omega = 39 \cdot 10^{-31}$ m^2/steradian and electron (hν=0.511 MeV)
$l_2 = 50$ mm
$S_2 = 10.0$ mm^2

Data inserted in Eq. (2.3.71) gives

$$\dot{\Phi}_s = \frac{3.0 \cdot 10^{10} \cdot 11.34 \cdot 10^3 \cdot 0.010 \cdot 0.003^2 \cdot \pi \cdot 2 \cdot 0.770 \cdot 10^{-2}}{4\pi \cdot 0.05^2}$$
$$\times \frac{6.022 \cdot 10^{26} \cdot 11.34 \cdot 10^3 \cdot 82 \cdot 1.0 \cdot 10^{-4} \cdot 10 \cdot 10^{-6} \cdot 39 \cdot 10^{-31}}{208 \cdot 0.05^2}$$

$\dot{\Phi}_s = 198$ m^{-2}s^{-1}

c) Photoelectric effect in the second disk.

Contribution of photons in the 80±10 keV range is obtained from a pair production or an incoherent scattering in the first disk followed by a photoelectric effect in the second disk. This will give K-x rays within the energy range 80±10 keV.

A) Pair production in the first disk and photoelectric effect in the second disk.

The fluence rate of fluorescence K-x rays is given by

$$\dot{\Phi}_{S_{pair,K}} = \frac{\dot{\Phi}_{sp}S_2(\tau_{K1}/\rho)\rho d_2\omega_K}{4\pi l_2^2} = \frac{\dot{\Phi}_{o}\rho d_1 S_1 2(\kappa/\rho)}{4\pi l_1^2}\frac{S_2(\tau_{K1}/\rho)\rho d_2\omega_K}{4\pi l_2^2} \qquad (2.3.72)$$

B) Incoherent scattering in the first disk and photoelectric effect in the second disk.

The fluence rate of fluorescence K-x rays is given by

$$\dot{\Phi}_{S_{incoh,K}} = \frac{\dot{\Phi}_{S_{incoh}}S_2(\tau_{K2}/\rho)\rho d_2\omega_K}{4\pi l_2^2} = \frac{\dot{\Phi}_{o}N_0 S_1 d_o\frac{d_e\sigma}{d\Omega}}{l_1^2}\frac{S_2(\tau_{K2}/\rho)\rho d_2\omega_K}{4\pi l_2^2} \qquad (2.3.73)$$

where $\tau_K/\rho = \tau/\rho \cdot (\tau_K/\tau)$ is the mass cross section for photoelectric effect in the K-shell and ω_K is the fluorescence yield for K-x rays.

Data:
$\tau_{K1}/\rho = 0.798 \cdot 10^{-2} \cdot 0.788\ m^2\ kg^{-1}\ (h\nu = 0.511\ MeV)$
$\tau_{K2}/\rho = 0.706 \cdot 10^{-2} \cdot 0.788\ m^2\ kg^{-1}\ (h\nu = 1.69\ MeV)$
$\omega_K = 0.967$
$d_e\sigma/d\Omega = 1.3 \cdot 10^{-31}\ m^2/steradian\ and\ electron\ (h\nu = 50\ MeV)$

Data inserted in Eq. (2.3.72) gives for A)

$$\dot{\Phi}_{S_{pair,K}} = \frac{3.0 \cdot 10^{10} \cdot 11.34 \cdot 10^3 \cdot 0.010 \cdot 0.003^2 \cdot \pi \cdot 2 \cdot 0.770 \cdot 10^{-2}}{4 \cdot \pi \cdot 0.05^2}$$
$$\times \frac{0.798 \cdot 10^{-2} \cdot 11.34 \cdot 10^3 \cdot 10 \cdot 10^{-6} \cdot 1.0 \cdot 10^{-4} \cdot 0.967 \cdot 0.788}{4 \cdot \pi \cdot 0.05^2}$$

Check of units:

$$\frac{kg \cdot m \cdot m^2 \cdot m^2}{m^2 \cdot s \cdot m^3 \cdot kg \cdot m^2}\frac{m^2 \cdot m^2 \cdot kg \cdot m}{kg \cdot m^3 m^2} = \frac{1}{m^2 \cdot s}$$

$\dot{\Phi}_{S_{pair,K}} = 1.035 \cdot 10^2\ m^{-2}s^{-1}$

Data inserted Eq. (2.3.73) gives for B)

$$\dot{\Phi}_{S_{incoh,K}} = \frac{3.0 \cdot 10^{10} \cdot 11.34 \cdot 10^3 \cdot \pi \cdot 0.003^2 \cdot 1.0 \cdot 10^{-2} \cdot 1.3 \cdot 10^{-31} \cdot 6.022 \cdot 10^{26} \cdot 82}{0.05^2 \cdot 208}$$

$$\times \frac{0.706 \cdot 10^{-2} \cdot 11.34 \cdot 10^3 \cdot 10 \cdot 10^{-6} \cdot 1.0 \cdot 10^{-4} \cdot 0.967 \cdot 0.788}{4 \cdot \pi \cdot 0.05^2}$$

Check of units:

$$\frac{electron \cdot m \cdot m^2 \cdot m^2}{m^2 \cdot s \cdot m^3 \cdot electron \cdot m^2} \frac{m^2 \cdot m^2 \cdot kg \cdot m}{kg \cdot m^3 m^2} = \frac{1}{m^2 \cdot s}$$

$\dot{\Phi}_{s_{incohK}} = 2.3 \ m^{-2} s^{-1}$

Total fluence rate for K-x rays is thus

$$\dot{\Phi}_s = 1.035 \cdot 10^2 + 2.3 = 1.06 \cdot 10^2 \ m^{-2} s^{-1}$$

During a time t the detector is hit by

$N = \dot{\Phi} t S_3$ (number of photons)

Data:
$t = 3600$ s (irradiation time)
$S_3 = 500 \ mm^2$ (detector area)

Data inserted gives the number of photons in the energy range 400±50 keV

$$N_{400} = 198 \cdot 3600 \cdot 500 \cdot 10^{-6} = 356$$

Number of photons in the energy range 80±10 keV

$$N_{80} = 1.06 \cdot 10^3 \cdot 3600 \cdot 500 \cdot 10^{-6} = 1.91 \cdot 10^2$$

Answer: Number of photons in the energy range 400±50 keV is 3.6·10^2 and the number of photons in the energy range 80±10 keV is 1.9·10^2

Solution exercise 2.24.
The fluence rate of the scattered photons at the detector is given by

$$\dot{\Phi}_s = \dot{\Phi}_0 \frac{mN_e \frac{d_e\sigma}{d\Omega}}{l_2^2} \tag{2.3.74}$$

where $\dot{\Phi}_s$ is the fluence rate of scattered photons at detector, $\dot{\Phi}_0$ is the fluence rate of primary photons at aluminum cylinder, $\frac{d_e\sigma}{d\Omega}$ is the Klein-Nishina cross section, N_e is the number of electrons per mass unit, m is the mass of the aluminum cylinder and l_2 is the the distance between the aluminum cylinder and the detector.

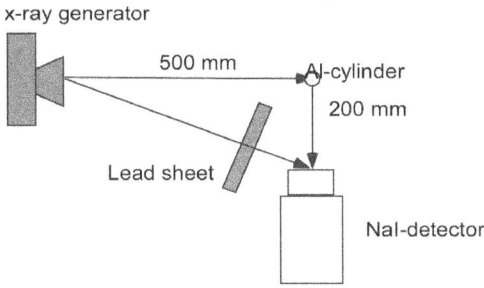

Figure 2.27: Irradiation geometry in exercise 2.24.

Fluence of transmitted photons through the lead shield is given by

$$\dot{\Phi}_t = \frac{\dot{\Phi}_0 l_1^2}{l_3^2} e^{-\mu_{Pb} d} \tag{2.3.75}$$

where $\dot{\Phi}_t$ is the fluence rate of transmitted photons at detector, l_1 is the distance between the x-ray source and the aluminum scatterer, μ_{Pb} is the linear attenuation coefficient in lead, d is the thickness of lead shield and l_3 is the distance between the x-ray source and the detector.

Data:
$\frac{d_e\sigma}{d\Omega} = 30 \cdot 10^{-31}$ m^2/steradian per electron (hν=80 keV)
$N_e = 6.022 \cdot 10^{26} \cdot 13/27$ (number of electrons per kg
$m = (0.5 \cdot 10^{-3})^2 \cdot \pi \cdot 12 \cdot 10^{-3} \cdot 2.7 \cdot 10^3$ kg
$l_1 = 500$ mm $= 0.50$ m
$l_2 = 200$ mm $= 0.20$ m
$\mu_{Pb} = 0.2417 \cdot 11.34 \cdot 10^3$ m^{-1} (hν=80 keV)
$l_3 = \sqrt{0.50^2 + 0.20^2} = \sqrt{0.29}$ m

Data inserted in Eq. (2.3.74) gives

$$\dot{\Phi}_s = \dot{\Phi}_0 \frac{(0.50 \cdot 10^{-3})^2 \cdot \pi \cdot 12 \cdot 10^{-3} \cdot 2.7 \cdot 10^3 \cdot 6.022 \cdot 10^{26} \cdot 13 \cdot 30 \cdot 10^{-31}}{27 \cdot 0.20^2}$$

$$= 5.534 \cdot 10^{-7} \dot{\Phi}_0$$

Check of units:

$$\frac{kg \cdot electron \cdot m^2}{m^2 \cdot s \cdot kg \cdot electron \cdot m^2} = \frac{1}{m^2 \cdot s}$$

Data inserted in Eq. (2.3.75) gives for the transmitted photons

$$\dot{\Phi}_t = \frac{\dot{\Phi}_0 \cdot 0.50^2 e^{-0.2417 \cdot 11.34 \cdot 10^3 \cdot d}}{(\sqrt{0.29})^2}$$

According to information

$$\dot{\Phi}_t = 0.02 \cdot \dot{\Phi}_s$$

Thus

$$\frac{\dot{\Phi}_0 \cdot 0.50^2 \cdot e^{-0.2417 \cdot 11.34 \cdot 10^3 \cdot d}}{(\sqrt{0.29})^2} = 0.02 \cdot 5.534 \cdot 10^{-7} \dot{\Phi}_0$$

Solving the equation gives

d=6.6 mm

Answer: The thickness of the lead shield needs to be 6.6 mm.

Solution exercise 2.25.

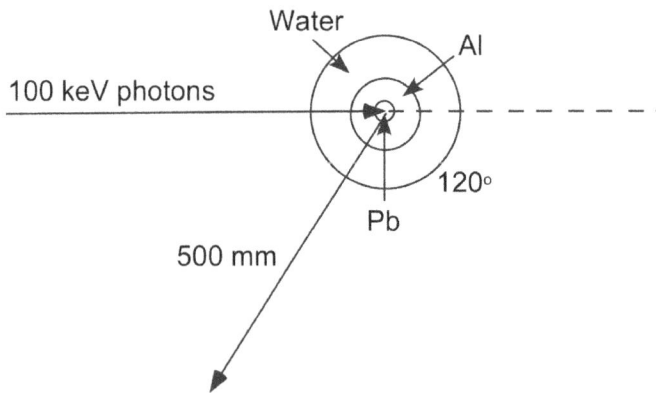

Figure 2.28: Simulation of the arm in exercise 2.25.

The possible interaction processes in the simulated arm are photoelectric effect or incoherent scattering.

The energy of incoherent scattered photons, with a primary energy of 100 keV is at a scattering angle of 120°

$$h\nu_s = \frac{100}{1 + \frac{100}{511}(1 - \cos 120)} = 77 \text{ keV} \tag{2.3.76}$$

This energy is not in the expected energy interval 80-90 keV.

Thus photons with an energy within the energy interval are K_β fluorescence x rays from the K-shell produced in Pb with the energy 85.5 keV (mean energy).

The photon fluence at the measuring point is given by

$$\Phi_s = \frac{\Phi_0(\tau_K/\rho)\omega_{K\beta}m_{Pb}}{4\pi l^2} e^{-\mu_{p,w}d_1} e^{-\mu_{p,Al}d_2} e^{-\mu_{p,Pb}d_3} e^{-\mu_{s,water}d_1} e^{-\mu_{s,Al}d_2} e^{-\mu_{s,Pb}d_3} \qquad (2.3.77)$$

where

$\Phi_0 = 4.5 \cdot 10^{12}$ m^{-2} (primary photon fluence at scatterer)

$m_{Pb} = 0.03 \cdot \pi \cdot 0.0005^2 \cdot 11.34 \cdot 10^3$ kg (mass of lead)

$\tau_K/\rho = (\tau_K/\tau)(\tau/\rho) = 0.788 \cdot 0.524$ m^2 kg^{-1} (cross-section for photoelectric effect in the K-shell (100 keV))

$\omega_{K\beta}=0.21$ (fluorescence yield for K$_\beta$-x rays in Pb)

$l=0.50$ m (distance between center of scatterer and measuring point)

$d_1=5.0 - 1.5=3.5$ cm$=0.035$ m (thickness of water)

$d_2=1.5 - 0.05=1.45$ cm$=0.0145$ m (thickness of aluminum)

$d_3=0.0005$ m (thickness of lead)

$\mu_{p,water} = 0.01711 \cdot 1.0 \cdot 10^3$ m^{-1} (linear attenuation coefficient in water for the primary photons (100 keV))

$\mu_{p,Al} = 0.01706 \cdot 2.7 \cdot 10^3$ m^{-1} (linear attenuation coefficient in Al for the primary photons (100 keV))

$\mu_{p,Pb} = 0.5552 \cdot 11.34 \cdot 10^3$ m^{-1} (linear attenuation coefficient in Pb for the primary photons (100 keV))

$\mu_{s,water} = 0.01796 \cdot 1.0 \cdot 10^3$ m^{-1} (linear attenuation coefficient in water for the secondary photons (85.5 keV))

$\mu_{s,Al} = 0.0191 \cdot 2.7 \cdot 10^3$ m^{-1} (linear attenuation coefficient in Al for the secondary photons (85.5 keV))

$\mu_{s,Pb} = 0.2051 \cdot 11.34 \cdot 10^3$ m^{-1} (linear attenuation coefficient in Pb for the secondary photons (85.5 keV))

$\rho_{water} = 1.0 \cdot 10^3$ kg/m^3, $\rho_{Al} = 2.7 \cdot 10^3$ kg/m^3, $\rho_{Pb} = 11.34 \cdot 10^3$ kg/m^3

Data inserted in Eq. (2.3.77) gives

$$\Phi_s = \frac{4.5 \cdot 10^{12} \cdot 0.788 \cdot 0.524 \cdot 0.21 \cdot 0.03 \cdot \pi \cdot 0.0005^2 \cdot 11.34 \cdot 10^3}{4 \cdot \pi \cdot 0.50^2}$$

$$\times e^{-.01711 \cdot 1.0 \cdot 10^3 \cdot 0.035} e^{-0.01706 \cdot 2.7 \cdot 10^3 \cdot 0.0145} e^{-0.5552 \cdot 11.34 \cdot 10^3 \cdot 0.0005}$$

$$\times e^{-0.01796 \cdot 1.0 \cdot 10^3 \cdot 0.035} e^{-0.0191 \cdot 2.7 \cdot 10^3 \cdot 0.0145} e^{-0.2051 \cdot 11.34 \cdot 10^3 \cdot 0.0005}$$

$\Phi_s=3.169 \cdot 10^4$ m^{-2}

Answer: The photon fluence in the energy range 80-90 keV is $3.2 \cdot 10^4$ m^{-2}.

Bibliography

Brahme A. (1975). Simple relations for the penetration of high energy electron beams in matter. SSI:19975-011. Internal report. Stockholm, Sweden: National Institute of Radiation protection.

Harder D. (1965). Energiespektren schneller Elektronen in vershiedenen Tiefen. In Zuppinger, A. and Puretti, G, (Eds). Symp on High energy Electrons (p.26-33), Montreaux. Berlin,Germany: Springer Verlag.

IAEA. (2000). Absorbed Dose Determination in External Beam Radiotherapy. Vienna, Austria. IAEA TRS-398.

3 Detectors and Measurements

3.1 Definitions and Relations

3.1.1 Counting Statistics

The Binomial Distribution

The most general statistical distribution is the Binomial distribution. The other distributions can be derived from this one. If n is the number of trials with a success probability of p, then the probability of obtaining x successful trials is given by

$$P(x) = \frac{n!}{(n-x)!x!} p^x (1-p)^{n-x} \tag{3.1.1}$$

$P(x)$ is a predicted probability distribution function defined only for integer values of n and x

The mean value of the distribution is

$$\bar{x} = pn \tag{3.1.2}$$

and the standard deviation is

$$\sigma = \sqrt{\bar{x}(1-p)} \tag{3.1.3}$$

The Poisson Distribution

If the success probability p is small and the number of trials n large then the Binomial distribution can be simplified to

$$P(x) = \frac{\bar{x}^x e^{-\bar{x}}}{x!} \tag{3.1.4}$$

where $\bar{x} = pn$. This is called the Poisson distribution.

Note that there is only one parameter, the mean value $\bar{x}$ needed to describe the Poisson distribution. The standard deviation is given by

$$\sigma = \sqrt{\bar{x}} \tag{3.1.5}$$

The Gaussian or Normal Distribution

If besides a low value of p, the mean value is large, the binomial distribution may be approximated by the Gaussian distribution

$$P(x) = \frac{1}{\sqrt{2\pi\sigma^2}} e^{\left(-\frac{(x-\bar{x})^2}{2\sigma^2}\right)} \tag{3.1.6}$$

If we use the information from the Poisson distribution that $\sigma^2 = \bar{x}$, then the Gaussian distribution may be written as

$$P(x) = \frac{1}{\sqrt{2\pi\bar{x}}}e^{\left(-\frac{(x-\bar{x})^2}{2\bar{x}}\right)}$$

(3.1.7)

This distribution is very useful as it is symmetric and continuous, contrary to the other distributions. It is also normalized like the other distributions

$$\int\limits_0^\infty P(x)dx = 1$$

(3.1.8)

This distribution is also described by only one single parameter $\bar{x} = np$ and the standard deviation is $\sigma = \sqrt{\bar{x}}$, as the Poisson distribution.

In general the standard deviation, σ, of a normal distribution can be independent of the mean value, $\bar{x}$, and then is a need for two parameters, σ and $\bar{x}$ to describe the function.

χ^2-test

To test if an obtained experimental distribution agrees with e.g. a Poisson distribution, or if there are some systematic errors, the χ^2-test can be used. The χ^2-value is given by

$$\chi^2 = \frac{1}{\bar{x}_e}\sum_{i=1}^N (x_i - \bar{x}_e)^2$$

(3.1.9)

where N is the number of measurements, $\bar{x}_e$ is the experimental mean value and x_i are the separate independent values.

For a Poisson distribution the χ^2-value should be close to $N - 1$. There are tables giving the probability that an obtained distribution is a Poisson distribution for a given χ^2-value, and this test is often of great value to check the measurements for systematic errors.

Uncertainty propagation

There are often several uncertainties in an experimental determination and these uncertainties will add up. If the uncertainties are individually small and symmetric around zero, the total uncertainty can be shown to be given by

$$\sigma_u^2 = \left(\frac{\delta u}{\delta x}\right)^2 \sigma_x^2 + \left(\frac{\delta u}{\delta y}\right)^2 \sigma_y^2 + \left(\frac{\delta u}{\delta z}\right)^2 \sigma_z^2 + \dots$$

(3.1.10)

where $u = u(x, y, z..)$ represents the derived quantity and σ_x, σ_y, σ_z,... are the standard deviations of the individual variables.

Using this relation for the standard variation of the net count rate one obtains

$$\sigma_{r_n} = \sqrt{\frac{r_T}{t_T} + \frac{r_b}{t_b}} = \sqrt{\frac{N_T}{t_T^2} + \frac{N_b}{t_b^2}} \tag{3.1.11}$$

where r_T is the count rate for sample including background, r_b is the count rate for background, N_T is the number of counts for sample including background, N_b is the number of counts for background, t_T is the calculation time for sample including background and t_b is the calculation time for background.

Rate meter statistics

The relative standard deviation of a rate meter is given by

$$\frac{\sigma_r}{r} = \frac{1}{\sqrt{2rRC}} \tag{3.1.12}$$

where r is the count rate and RC is the time constant.

The equilibrium time, t_0, of a rate meter is defined as the time that is needed to obtain a value that is equal to the saturation value subtracted with one standard deviation. This is obtained from the relation

$$t_0 = \frac{RC}{2} \ln(r2RC). \tag{3.1.13}$$

3.1.2 Detector Properties

Full width half maximum

The energy resolution of a detector is often described by the full width half maximum $w_{1/2}$. If it is assumed that the width of the energy peak is due to statistical variation following a Poisson distribution and that this may be approximated by a Gaussian distribution, $w_{1/2}$ is given by

$$w_{1/2} = 2.35E\sqrt{\frac{F}{N}} = 2.35\sqrt{EF\epsilon} \tag{3.1.14}$$

where F is the Fano factor (the ratio between the observed variance and the Poisson expected variance), N is the number of charge carriers, E is the peak energy and ϵ is the energy to produce a charge carrier.

Measurement geometry for a point source on the central axis

The geometry factor, G, is defined as the number of particles that are emitted by the source and hitting the detector. For a point source positioned at the central axis of a cylindrical detector, see Fig. 3.1, G is given by

$$G = \frac{1}{2}(1 - \frac{h}{\sqrt{h^2 + a^2}}) = \frac{1}{2}(1 - \cos(\theta/2)) \tag{3.1.15}$$

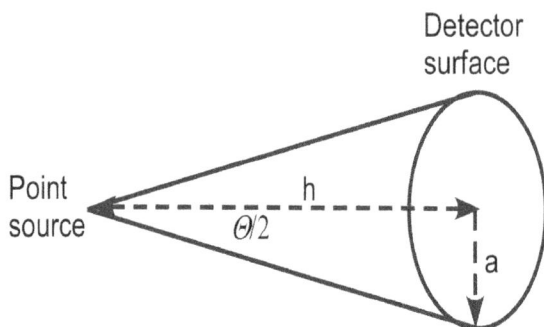

Figure 3.1: Geometry factor for a point source.

where h is the distance source-detector surface, a is the detector radius and $\theta/2$ is half of the opening angle.

Note that this definition is different from the solid angle Ω, which for a cylindrical detector surface is given by

$$\Omega = 2\pi(1 - \frac{h}{\sqrt{h^2 + a^2}}) \qquad (3.1.16)$$

Thus

$$G = \frac{\Omega}{4\pi} \qquad (3.1.17)$$

When the source to surface distance is large compared to the detector radius, the geometry factor can be simplified to

$$G = \frac{a^2}{4h^2} \qquad (3.1.18)$$

and the solid angle, Ω, to

$$\Omega = \frac{\pi a^2}{h^2} \qquad (3.1.19)$$

Correction for resolving time

Pulse detectors and the electronic measuring devices need a finite time to register a pulse. The minimum time between two separate pulses is called the resolving time or the dead time of the counter. This dead time may result in a reduction in count rate. The correction for this is dependent if the counter is paralyzable or non-paralyzable. A paralyzable detector does not register a pulse during the dead time, but the dead time is prolonged. This prolongation will not occur for non-paralyzable counters. The equations for the two different detectors are given as:

Non paralyzable counters

$$r_0 = \frac{r}{1 - r\tau} \qquad (3.1.20)$$

Paralyzable counters

$$r = r_0 e^{-r_0 \tau} \tag{3.1.21}$$

where r_0 is the corrected counting rate, r is the measured counting rate and τ is the resolving time (dead time) of the counter.

Coincidence measurements

If two particles from the same decay are absorbed by a detector their energy may be added and give rise to a "sum-peak". The true counting rate in the sum peak is given by

$$r_{12} = A \eta_1 \eta_2 f_1 f_2 G^2 \tag{3.1.22}$$

where A is the source activity, η_1, η_2 detector efficiency for particle 1 and particle 2 respectively, f_1, f_2 are the number of particles per decay for particle 1 and particle 2 respectively and G is the geometry factor.

The detector will register two pulses as one within a certain resolving time, τ. This means that beside the true counting coincidence rate, there is a possibility that two particles from different decays hit the detector at nearly the same time and will be detected as a coincidence count. This random coincidence rate is given by

$$r_{ch} = r_1 r_2 \tau = A^2 \eta_1 \eta_2 G^2 f_1 f_2 \tau \tag{3.1.23}$$

Sometimes two detectors are used in coincidence. In a positron camera the annihilation photons are emitted in coincidence and can be measured by two detectors in opposite direction from the point of annihilation. It is also often possible to reduce background count rates by using two detectors in coincidence. Beside the true coincidences there is also here a possibility that two particles from different decays are counted. This random coincidence rate is approximately given by

$$r_{ch} = 2 r_1 r_2 \tau_c \tag{3.1.24}$$

where r_1, r_2 are the counting rates of the separate detector and τ_c the resolving time of the coincidence circuit.

Detector relations
A. Gas detectors

Gas multiplication in a proportional counter

$$\ln M = \frac{V}{\ln(r_2/r_1)} \frac{\ln 2}{\Delta V} \left(\ln \frac{V}{p r_1 \ln(r_2/r_1)} - \ln K \right) \tag{3.1.25}$$

where M is the gas multiplication factor, V is the applied voltage, r_1 is the anode radius, r_2 is the cathode radius, p is the gas pressure, and K is the Diethorn-parameter for the proportional counter.

Relative standard deviation in the charge

$$\left(\frac{\sigma_Q}{Q}\right)^2 = \left(\frac{\sigma_{n_0}}{n_0}\right)^2 + \frac{1}{n_0}\left(\frac{\sigma_A}{\bar{A}}\right)^2 \tag{3.1.26}$$

$$\left(\frac{\sigma_A}{\bar{A}}\right)^2 \approx 0.5 \text{ if } \bar{A} \text{ is large} \tag{3.1.27}$$

where A is the electron multiplication factor, $\bar{A}$ is the mean value of A and n_0 is the number of produced ion pairs.

Pulse drift in a cylindrical gas detector

The velocity of the ions at radius r is given by

$$v^+(r) = \frac{\mu}{p}\frac{V_0}{\ln(r_2/r_1)}\frac{1}{r} \tag{3.1.28}$$

The position of the ions at time t is given by

$$r(t) = \sqrt{\frac{2\mu}{p}\frac{V_0 t}{\ln(r_2/r_1)} + r_1^2} \tag{3.1.29}$$

The time needed to collect the ions is given by

$$t^+ = \frac{(r_2^2 - r_1^2)p\ln(r_2/r_1)}{2\mu V_0} \tag{3.1.30}$$

The variation of the detector signal with time is given by

$$V_r(t) = \frac{Q}{C}\frac{1}{\ln(r_2/r_1)}\ln\left(\frac{2\mu V_0 t}{r_1^2 p\ln(r_2/r_1)} + 1\right)^{1/2} \tag{3.1.31}$$

where r_1 is the anode radius, r_2 is the cathode radius, p is the pressure in the chamber, V_0 is the voltage over the chamber, μ is the mobility, C is the detector capacitance, and Q is the collected charge.

B. Scintillation detectors

Pulse shape from a photomultiplier tube is given by

$$V(t) = \frac{1}{\lambda - \theta}\frac{\lambda Q}{C}(e^{-\theta t} - e^{-\lambda t}) \tag{3.1.32}$$

where λ is the scintillation detector decay time, θ is the anode time constant, Q is the total charge, and C is the anode capacitance

With a large time constant $(\theta \ll \lambda)$

$$V(t) = \frac{Q}{C}(e^{-\theta t} - e^{-\lambda t})$$

(3.1.33)

With a small time constant $(\theta \gg \lambda)$

$$V(t) = \frac{\lambda}{\theta}\frac{Q}{C}(e^{-\lambda t} - e^{-\theta t})$$

(3.1.34)

C. Semiconductor detectors

The depletion depth d is given by

$$d \cong \sqrt{\left(\frac{2\epsilon V}{eN}\right)}$$

(3.1.35)

where ϵ is the dielectric constant of the material, N is the concentration of doped atoms and V is the applied voltage.

D. Activation detector

The saturation activity in activation of a material is given by

$$A_\infty = \frac{\lambda(C - B)}{\eta(1 - e^{-\lambda t_0})(e^{-\lambda t_1} - e^{-\lambda t_2})}$$

(3.1.36)

where C is the total number of pulses measured during time t_1 to t_2, time measured from the end of irradiation. B is the number of background pulses during the same time, t_0 is the irradiation time, λ is the decay constant and η is the detector efficiency.

E. Neutron scintillation detector.

The detector efficiency for an organic scintillation detector made of carbon and hydrogen irradiated with neutrons is given by

$$\eta = \frac{N_H \sigma_H}{N_H \sigma_H + N_C \sigma_C}\left(1 - e^{-(N_H \sigma_H + N_C \sigma_C)d}\right)$$

(3.1.37)

where N_H, N_C is the number of target nuclides per volume unit in hydrogen and carbon respectively. σ_H, σ_C is the reaction cross section in hydrogen and carbon respectively. d is the detector thickness.

3.2 Exercises in Detectors and Measurements

3.2.1 Counting Statistics

Exercise 3.1. Derive an expression that optimizes data acquisition lengths between measurement of a sample plus background and background alone, respectively, for a specified total acquisition length. Then calculate the optimal acquisition time for sample with background and background alone, when given a total acquisition time of 0.50 h and count rates of 100 and 20 min^{-1}, respectively. Calculate also the relative standard deviation of the net count rate.

Exercise 3.2. The current from an ion-chamber exposed to γ-radiation is $1.0 \cdot 10^{-13}$ A. The output from the electrometer is connected to a printing system. What will the expected relative standard deviation of the printout be and what will the equilibrium time be, knowing that the system time constant $\tau = 2.0$ s and the average path-length of secondary electrons under radiation equilibrium is 0.20 m air at NTP? Assume that the energy loss of secondary electrons through collision processes equals 0.20 MeV m^2 kg^{-1}.

Exercise 3.3. A radiation detector is measuring at the sarcophagus at Chernobyl to check possible emission of radioactivity. The mean dose rate is 12.5 μGy h^{-1}, which corresponds to a current in the detector of 7.5 nA. The time constant of the detector is 5.0 s. The detector shall alarm if the dose rate is increasing with more than 3.0 standard deviations. Which dose rate is this corresponding to? Every pulse in the detector has a charge of 0.19 nC.

Exercise 3.4. A dairy uses a scintillation detector to check if the activity concentration of ^{137}Cs in the milk is lower than 50 Bq l^{-1}. The staff at the dairy is not used with this type of measurements and they have a feeling that the background is higher if there is a person present in the detector room. The following result as shown in the table was obtained. The detector measured photon energies in the energy range 600-700 keV.

Sample	Measuring time	Number of counts
Milk sample (10 ml)	20.0 min	456
Background (without person)	40.0 min	210
Background (with person)	30.0 min	165

The efficiency of the detector in the measuring region was determined to be 0.62±0.02 counts (1.0 SD) per photon in the energy range 600-700 keV.

a) Determine the standard deviation of the difference in background count rate with or without a person. Discuss if the difference in background count rate with or without a person is significant?

b) Calculate the concentration of the activity in $\mathrm{Bq\,l^{-1}}$ and the standard deviation of the concentration. Should one consider the milk as radioactive according to the limit $50\,\mathrm{Bq\,l^{-1}}$?

Exercise 3.5. You are responsible of an air sample unit, placed outside a plant using radioactive nuclides. There is a suspicion of a leakage of $^{239}\mathrm{Pu}$. The air sample unit has an air sampling rate of $1.56\,\mathrm{m^3\,min^{-1}}$ and the air filter has an efficiency of 0.80. The air sampling time is 60.0 min.

After the sampling, there is a delay period of 24.0 h, resulting in that all natural radioactive nuclides from Rn have decayed. Then the filter is measured during 600 s and 220 counts are registered. Calculate the activity concentration of $^{239}\mathrm{Pu}$ in $\mathrm{Bq\,m^{-3}}$ and the standard deviation in the determination of the concentration. Only counting statistics need to be considered.

The self absorption of the alpha-particles in the filter implies that only 40% of the emitted alpha-particles leave the filter in the direction of the detector and can be detected. The filter is placed close to an alpha detector with an active detector area of $60.0\,\mathrm{cm^2}$. The background is 20.0 counts during 100 min. The detector efficiency for alpha-particles that hit the detector is 0.30 and the area of the active filter is $500\,\mathrm{cm^2}$.

Exercise 3.6. The half life of a radioactive source shall be determined. The radiation source is measured twice, each time with the measuring time 600 s and with 24 h between the measurements. In the first measurement 1683 counts are obtained and in the second 914 counts, both including background counts. The count rate for the background is $21.0\,\mathrm{min^{-1}}$, determined with a measurement time of 20.0 min. Determine the half life of the source and the uncertainty in the determination.

Exercise 3.7. The linear attenuation coefficient for an unknown material shall be determined. The following data are obtained in a measurement made in "narrow beam geometry".

Material thickness: $2.53\pm0.02\,\mathrm{cm}$ (1 standard deviation).
Measurement without material: 35 000 counts during 300 s (including background).
Measurement with material: 25 700 counts during 300 s (including background).
Measurement of the background (both measurements): 2350 counts during 600 s.
The uncertainty in the measurement time can be neglected.

Determine the linear attenuation coefficient and the standard deviation.

Exercise 3.8. At a company that produces ethanol, there is a suspicion that somebody steals the ethanol and replaces it with water. To check this without a need to open the closed bottles, they are irradiated with γ-rays from a ^{137}Cs-source with an activity of 5.0 MBq, in a collimated beam with the diameter at the detector of 20 mm. This can be regarded as a narrow beam geometry. The distance between the source and the detector is 60.0 cm. A NaI-detector is used for the measurements. The diameter of the detector is 50.0 mm and the intrinsic efficiency for the used photon energy is 0.73. The transmission through a bottle filled with ethanol is compared with the transmission through a bottle with suspected content. The bottles are square shaped with a thickness of the content of 80.0 mm. Determine the measurement time for each bottle in order to be able to obtain a difference in count rate of 2.0 standard deviations between a bottle with water and a bottle with ethanol. Neglect the attenuation in the wall of the bottle. The density of ethanol is $0.791 \cdot 10^3$ kg m^{-3}. The count rate for the background has been determined to be 17.0 s^{-1} during a measurement time of 100 s. The correction for dead time may be neglected.

Exercise 3.9. The activity of a point source placed 0.50 m below the ground level is to be determined by the use of a detector located on the ground level, facing downwards. The area of the detector is 10.0 cm^2. The intrinsic efficiency of the detector is determined to 0.60 ± 0.01. The build-up factor of the secondary photons is 1.10 ± 0.1. The linear attenuation coefficient of ground material is 2.83 ± 0.050 m^{-1}. A total of 10 000 counts is achieved during 60.0 s. The number of background counts is $1.20 \cdot 10^4$ during 20 min. Determine the activity of the source and the uncertainty in the activity assuming one emitted photon per decay. Neglect the uncertainty in the distance and the detector diameter and assume that given uncertainties are one standard deviation.

Exercise 3.10. The activity of a ^{137}Cs-source is going to be determined with a scintillation detector. The detector has a diameter of 5.08 cm. The radiation source can be positioned at an effective distance of 20 or 30 cm from the detector. The standard deviation in the determination of the distance is estimated to be 0.50 cm. The radiation source is small and can be regarded as a point source. The efficiency of the detector for the γ-radiation from ^{137}Cs is 0.76. When positioning the source at an effective distance of 30 cm, 3562 counts are obtained during 60.0 s. The number of background counts is 3826 obtained during 600 s. Calculate the activity of the source and the standard deviation in the activity if only counting statistics and the uncertainty in the distance are included. The approximate estimation of the geometry factor may be used. Neglect the attenuation in the air.

If the radioactive source instead is positioned at an effective distance of 20 cm and the measurement is made with the same measuring time, will this result in a smaller or larger uncertainty?

3.2.2 Detector Properties

Exercise 3.11. In a measuring set up, the average count rate r of a radioactive source of ^{11}C is determined by measuring the number of counts during 30.0 min. The measurement is regarded as representative of the activity of the sample after 15.0 min, i.e. at the center of the counting period. This value will however differ from the real count rate at this point of time. Calculate the ratio of the two count rates.

Exercise 3.12. The possible angular dependence of the two photons emitted from a ^{60}Co-source shall be determined. Two NaI-scintillation detectors are positioned according to Fig. 3.2 and the coincidence rate of the two detectors is measured. The pulse height analyzer is open for energies between 1.0 and 1.5 MeV. The efficiency of the detectors is 0.45 in this energy range. Which coincidence rate is expected if there is no angular dependence between the two gamma-rays? The resolving time for the coincidence circuit is 2.0 μs and the coincidence background count rate is $2.0\,\mathrm{s}^{-1}$. The activity of the radioactive source is 3.90 MBq. The diameter of the scintillation detector is 50 mm.

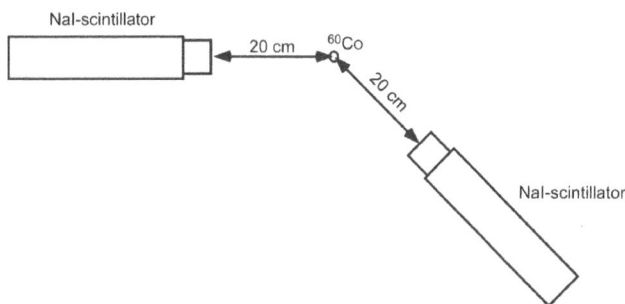

Figure 3.2: Geometry of the coincidence measurement in exercise 3.12.

Exercise 3.13. A radionuclide is emitting two coincident gamma photons with 100% exchange per decay. A sample of this radionuclide is placed at the central axis 10.0 cm from the surface of a cylindrical detector with a radius of 5.0 cm. The efficiency in the total-absorption peak is 50% for one of the photons and 30% for the other one. The total count rate in the sum peak is $129\,\mathrm{s}^{-1}$. Calculate the activity of the source if the detector has a resolving time of 2.0 μs, and the background count rate in the sum peak is $12\,\mathrm{s}^{-1}$.

Exercise 3.14. Calculate the largest dead time for a GM-counter in order to have dead time losses less than 1.0%, if the detector is used to measure the count rate of a ^{14}C radioactive source with the activity 3.0 kBq. The radioactive source can be regarded as a point and is placed at the central axis at a distance of 40 mm from the detector surface. The effective diameter of the detector is 30 mm. The transmission through the detector window is 0.87. The counting efficiency of the particles that pass through the window is 1.0.

Exercise 3.15. A cylindrical GM-counter with a diameter of 40 mm and an anode diameter of 0.16 mm is filled with argon and ethyl alcohol. The mean free path of the electrons between two collisions at the pressure 101.3 kPa and temperature 20°C is 6.4·10^{-4} mm. The maximal radius at which secondary electrons can be produced is 0.65 mm. The voltage over the tube is 1500 V. The ionization energy is 15.7 eV. Calculate the gas pressure in the tube if the temperature is 20°C.

Exercise 3.16. A cylindrical gas detector has a central anode with a radius of 25 μm and a cathode with a radius of 25 mm. The voltage over the tube is 1000 V. At what distance from the center will an electron get enough energy, when traveling a mean free path length of 1.0 mm, to be able to ionize helium, which has an ionization energy of 23 eV?

Exercise 3.17. A GM-counter has an anode radius of 20 μm and a cathode radius of 20.0 mm. The pressure in the chamber is 10.7 kPa. If the voltage over the counter is 600 V, how far have the positive ions traveled when the pulse amplitude has increased to 25% of its maximum? The time constant of the external circuit is assumed to be long compared to the collection time. The ions are assumed to be produced at the anode. Their mobility is 1.40·10^4 (mm s^{-1})(V mm^{-1})$^{-1}$kPa.

Exercise 3.18. A cylindrical gas detector with a cathode radius of 20 mm and an anode radius of 0.50 mm can theoretically be used both as an ionization chamber and as a GM-counter. A ^{59}Fe point radiation source with the activity 2.76 MBq is positioned centrally at a distance of 50 mm from the tube end window. The window has a thickness of 1.5 mg cm^{-2} and is made of Mica. The electrons emitted from the source are assumed to be totally absorbed in the detector, but the contribution from the photons may be neglected. The absorption of the electrons in the window may also be neglected. The gas in the detector is argon with a mean energy to produce an ion pair, $\bar{W}$=35.0 eV.

When the voltage over the chamber is low the detector is working as an ionization chamber. Calculate the produced ionization current.

The voltage is then increased and the detector is now working as a GM-counter. Which voltage is necessary if the maximal radius for which secondary electrons can be produced is 0.65 mm? The ionization energy is 15.7 eV and the mean free path of the electrons at 101.3 kPa at room temperature is 6.4·10^{-4} mm. The pressure in the chamber is 1.50 kPa.

How many counts per second are expected if the dead time is 130 μs?

Exercise 3.19. In a plane parallel ionization chamber the distance between the electrodes is 2.0 mm. In order to have a fast collection of the ions it is necessary to have a high field strength. However, if the voltage is too high, there is a risk of increased charge leakage and flash-over. The maximum practical value of the voltage is then estimated to 400 V. Calculate the maximal and the minimal collection time for this voltage. Assume that the liberated electrons do not produce negative ions. Calculate also the pulse voltage obtained if an electron with the energy 0.7 MeV is absorbed in the chamber, and the produced charge is measured over a capacitance of 13 pF.

The ionization chamber is filled with air. The temperature and the pressure in the chamber is 21°C and 100 kPa, respectively. The mobility of the ions is $1.32 \cdot 10^4$ (mm s^{-1})(V mm^{-1})$^{-1}$kPa and $1.62 \cdot 10^7$ (mm s^{-1})(V mm^{-1})$^{-1}$kPa for electrons.

Exercise 3.20. Calculate the necessary amplification that is needed in order that the most energetic photon from a ^{60}Co radioactive source that is totally absorbed in a NaI-scintillator, will give a pulse in channel 500 in a multichannel analyzer, where channel 1024 corresponds to a voltage of 10.0 V. The following data are given for the detector:
Light conversion efficiency=12%.
Light photon wave length=415 nm.
Light collection efficiency=50%.
Photocathode efficiency=20%.

80% of the produced electrons in the photocathode are collected at the first dynode in the PM-tube. The PM-tube has 10 dynodes, with a multiplication factor of δ=2.5. The anode has a load resistor of 100 kΩ and a capacitance of 100 pF. The decay constant of the NaI-scintillator is $4.35 \cdot 10^6$ s^{-1}.

Exercise 3.21. A HpGe-detector is used to detect 90° Compton scattered radiation. A voltage pulse of 1.25 V is obtained after the preamplifier, which has an amplification of 1500 and an input capacitance of 7.5 pF. If the scattered photon is totally absorbed in the HpGe-detector, what is the energy of the primary photon? If the energy spread in the total absorption peak only depends on statistics in the electron-hole production what will the FWHM (full width half maximum) be? The Fano factor is 0.10.

Exercise 3.22. A NaI-scintillator connected to a rate meter is used to check if a material contains ^{65}Zn. Calculate the smallest activity that can be determined with the following conditions.

The NaI-scintillator has a thickness of 50 mm. The radiation source is considered to be a point source centrally positioned at a distance of 200 mm from the surface of the detector. In front of the detector a collimator with a diameter opening of 20 mm is positioned. This means that for calculation of absorbed energy it is possible to assume a parallel impinging beam of photons. Neglect escape of characteristic x rays. Assume that Compton scattered photons are not absorbed in the scintillator. To produce a photoelectron at the photocathode 100 eV is needed on average. Assume that all photoelectrons produced at the photocathode reach the first dynode. The PM-tube has 10 dynodes with a multiplication factor of 3.5. There is a background current of 124 µA. The size of the pulses from the background radiation is $2.39 \cdot 10^{-11}$ C. The time constant of the rate meter is 20.0 s. The measured total current shall differ from the background current with 2.0 standard deviations of the background current. Assume that the decay of the pulses can be neglected.

Exercise 3.23. In radiotherapy with high photon energies, there will be (γ,n) reactions that will produce a fluence rate of neutrons in the treatment room. To determine the fluence of thermal neutrons, ^{115}In-foils are placed at different positions in the room. The irradiation time is 10.0 min. A measurement of an ^{116m}In-foil activity starts 10.0 minutes after finishing the irradiation. The measurement time is 20.0 min. A total amount of 76 800 counts is obtained. The background count rate have been determined to 13 s^{-1}. Calculate the neutron fluence rate, if the absolute efficiency of the detector is 0.30 pulses per photon and the number of photons per decay is 0.48 in the relevant measurement region. The activation cross section for ^{115}In for the energy range of interest is $2640 \cdot 10^{-28}$ m^2. The mass of the foil is 2.0 g.

Exercise 3.24. There is an interest to determine how much silver there is in an old coin. The coin is irradiated with a fluence of thermal neutrons with a fluence rate of $10.2 \cdot 10^7$ m^{-2}s^{-1}. The coin is irradiated during 600 s and then the electrons, which are emitted from the activated coin, are measured during a 300 s period, with a break of 60 s between the irradiation and the measurement. A total amount of 25 000 counts is obtained. The background count rate has been determined to 3.0 s^{-1}. Calculate the mass of silver in the coin. Neglect the self absorption of the electrons in the coin and assume that the detector efficiency for electrons is 0.10 in the geometry used. The activation cross sections per nucleus are: $\sigma(^{107}$Ag,^{108}Ag$)=30 \cdot 10^{28}$ m^2, $\sigma(^{109}$Ag,^{110}Ag$)=110 \cdot 10^{28}$ m^2. Natural silver contains 51.35% ^{107}Ag and 48.65% ^{109}Ag.

Exercise 3.25. A spherical proportional counter "without walls" and with the diameter 10 mm is placed in a vacuum tank. This is filled with tissue equivalent gas according to Rossi (64.4% CH$_4$, 32.4% CO$_2$, 3.2% N$_2$) at a pressure of 9.44 kPa. The detector is irradiated with α-particles from ^{241}Am in a parallel beam. The fluence rate is very low and each event is registered separately. The gas multiplication is $1.0 \cdot 10^3$. The capacitance of the preamplifier is 1.0 pF and the signal is then amplified 10 times before it is analysed in a multichannel analyser with 1024 channels and with the pulse height interval 0-10 V. Determine the shape of the spectrum. The measurement is made at 293 K. The density of the gas at 273 K and 101.3 kPa is 1.128 kg m^{-3}. $\bar{W}$=31.1 eV (mean energy to produce an ion pair).

3.3 Solutions in Detectors and Measurements

3.3.1 Counting Statistics

Solution exercise 3.1.
The standard deviation of the net counting rate is given by

$$s_n = \sqrt{\frac{r_T}{t_T} + \frac{r_b}{t_b}} \tag{3.3.1}$$

where r_T=count rate for sample including background, r_b=count rate for background, t_T=calculation time for sample including background and t_b=calculation time for background. Assume that the total time for measurement is $\tau = t_T + t_b$.

Inserting this in Eq. (3.3.1) and squaring the result, the following relation is obtained.

$$s_n^2 = \frac{r_T}{t_T} + \frac{r_b}{\tau - t_T} \tag{3.3.2}$$

Derivation gives

$$\frac{d(s_n^2)}{dt_T} = -\frac{r_T}{t_T^2} + \frac{r_b}{(\tau - t_T)^2} \tag{3.3.3}$$

Reorganizing Eq. (3.3.3) and setting $d(s_n^2)/dt_T$ equal to zero gives the optimal distribution of measurement times for a minimal variance or standard deviation.

$$\frac{r_T}{t_T^2} = \frac{r_b}{(\tau - t_T)^2} \tag{3.3.4}$$

and

$$\frac{t_T}{t_b} = \sqrt{\frac{r_T}{r_b}} \tag{3.3.5}$$

Use this relation for the following data:

τ=30 min, r_T=100 min^{-1}, r_b=20 min^{-1}

This inserted in Eq. (3.3.5) gives

$$\frac{t_T}{30 - t_T} = \sqrt{\frac{100}{20}}$$

This gives t_T=20.7 min and t_b=30-20.7=9.3 min

The relative standard deviation is then

$$\frac{s_n}{r_n} = \frac{\sqrt{\frac{100}{20.7} + \frac{20}{9.3}}}{100 - 20} = 0.033$$

Answer: The optimal measurement times are 20.7 min for sample including background and 9.3 min for background only. The relative standard deviation in the net count rate is 3.3%.

Solution exercise 3.2.

The number of ion pairs, N, produced by a passage of a secondary electron is given by

$$N = \frac{d\rho(\frac{dE}{\rho \, dl})}{\bar{W}} \tag{3.3.6}$$

The charge per pulse is $q = Ne$. The measured current is $I = rq$. Thus the count rate is given by the equation

$$r = \frac{I}{q} = \frac{I\bar{W}}{d\rho(\frac{dE}{\rho \, dl})e} \tag{3.3.7}$$

where
d=0.20 m (average path length of the secondary electron)
ρ=1.293 kg m^{-3} (density of air at NTP)
$(dE/\rho \, dl)$=0.20 MeV m^2 kg^{-1} (mass stopping power of the secondary electrons)
$\bar{W}$=33.97 eV (mean energy to ionize an ion pair in air)
e=1.602·10^{-19} C (charge of an electron)
I=1.0·10^{-13} A (measured current)
r=count rate

Data inserted in Eq. (3.3.7) gives

$$r = \frac{1.0 \cdot 10^{-13} \cdot 33.97}{0.20 \cdot 0.20 \cdot 10^6 \cdot 1.293 \cdot 1.602 \cdot 10^{-19}} = 410 \, \text{s}^{-1}$$

The standard deviation in the count rate is given by

$$\frac{s_r}{r} = \frac{1}{\sqrt{r2RC}} \tag{3.3.8}$$

where RC=2.0 s (time constant of the system).

Data inserted gives

$$\frac{s_r}{r} = \frac{1}{\sqrt{410 \cdot 2 \cdot 2}} = 2.47 \cdot 10^{-2}$$

The equilibrium time, t_0, of the printer is obtained from the relation

$$t_0 = \frac{RC}{2} \ln(r2RC) \tag{3.3.9}$$

Data inserted gives

$$t_0 = \frac{2}{2} \ln(410 \cdot 2 \cdot 2) = 7.4 \text{ s}$$

Answer: The relative standard deviation is 2.5% and the equilibrium time is 7.4 s.

Solution exercise 3.3.
The standard deviation of a rate meter is given by

$$s_r = \frac{r}{\sqrt{r2RC}} \tag{3.3.10}$$

where RC is the time constant of the detector (5.0 s) and r is the count rate.

The count rate is given by the relation $r=I/q$ where

$I = 7.50 \cdot 10^{-9}$ A (measured current)
$q = 0.19 \cdot 10^{-9}$ C (charge per pulse)

Thus

$$s_r = \frac{(7.5 \cdot 10^{-9}/0.19 \cdot 10^{-9})}{\sqrt{2 \cdot (7.5 \cdot 10^{-9}/0.19 \cdot 10^{-9}) \cdot 5}} = 1.987 \text{ s}^{-1} \tag{3.3.11}$$

Three standard deviations are then 5.96 s^{-1}. This corresponds to an increase in current of

$\Delta I = 5.96 \cdot 0.19 \cdot 10^{-9} = 1.13$ nA.

This means that the current after increase should be 7.50+1.13=8.63 nA.

This corresponds to a dose rate of 8.63·(12.5/7.50)=14.4 µGy h^{-1}.

Answer: The dose rate after the increase is 14.4 µGy h^{-1}.

Solution exercise 3.4.
The difference in the background count rate in the measurements with and without a person in the room is given by

$$\Delta r_b = \frac{N_2}{t_2} - \frac{N_1}{t_1} \tag{3.3.12}$$

The standard deviation in the difference is given by

$$s_{\Delta r_b} = \sqrt{\frac{N_1}{t_1^2} + \frac{N_2}{t_2^2}} \tag{3.3.13}$$

where
N_1=210 (number of background counts without a person)
N_2=165 (number of background counts with a person)
t_1=40 min (time for measurement of background counts without a person)
t_2=30 min (time for measurement of background counts with a person)

Data inserted gives

$$\Delta r_b = \frac{165}{30} - \frac{210}{40} = 0.25 \, \text{min}^{-1}$$

and

$$s_{\Delta r_b} = \sqrt{\frac{165}{30^2} + \frac{210}{40^2}} = 0.56 \, \text{min}^{-1}$$

The difference Δr_b is less than half of one standard deviation. It is unlikely that there is a difference between the two background counting measurements. The total background is used in the further calculations.

The activity of the sample is given by

$$A = \frac{r_n}{\eta f} \tag{3.3.14}$$

where
η=0.62±0.02 (efficiency of the detector)
f=0.946·0.898 (number of 0.662 keV photons per decay)

The net count rate, r_n, is

$$r_n = \frac{456}{20} - \frac{375}{70} = 17.44 \, \text{min}^{-1}$$

The standard deviation in r_n is

$$s_{r_n} = \sqrt{\frac{456}{20^2} + \frac{375}{70^2}} = 1.029 \, \text{min}^{-1}$$

Data inserted in Eq. (3.3.14) gives the activity for a 10 ml sample

$$A = \frac{17.44}{60 \cdot 0.62 \cdot 0.946 \cdot 0.898} = 0.55 \, \text{Bq}$$

Thus the activity concentration C is $55\,\mathrm{Bq}^{-1}$.

The standard deviation of the activity concentration is given by the relation

$$s_C = C\sqrt{(\frac{1}{r_n})^2 s_{r_n}^2 + (\frac{1}{\eta})^2 s_\eta^2} \tag{3.3.15}$$

Data inserted in Eq. (3.3.15) gives

$$s_C = 55\sqrt{\left(\frac{1.03}{17.44}\right)^2 + \left(\frac{0.02}{0.62}\right)^2} = 3.90\,\mathrm{Bq}^{-1}$$

This means that the concentration of ^{137}Cs-activity is larger than $50\,\mathrm{Bq}^{-1}$ with slightly more than one standard deviation. The milk should be considered as radioactive.

Answer: The standard deviation in the difference in the two background measurements is $0.56\,\mathrm{min}^{-1}$. The concentration of ^{137}Cs-activity is $55\pm4\,\mathrm{Bq}^{-1}$ (One standard deviation).

Solution exercise 3.5.
The net count rate r_n of the detector is given by

$$r_n = \frac{C\eta_d\eta_f S_A F t S_d f}{S_f} \tag{3.3.16}$$

The concentration of ^{239}Pu, C, is then

$$C = \frac{r_n S_f}{\eta_d\eta_f S_A F t S_d f} \tag{3.3.17}$$

where

$$r_n = \frac{N_f}{t_f} - \frac{N_b}{t_b} \tag{3.3.18}$$

Data:
N_f=220 (number of counts with filter)
t_f=600 s (time for filter measurement)
N_b=20 (number of counts of background)
t_b=6000 s (time for background measurement)
S_f=500 cm^2 (filter area)
η_d=0.30 (detector efficiency)
η_f=0.80 (filtration efficiency)
S_A=0.40 (self absorption in the filter)
F=1.56 m^3 min^{-1} (air sampling rate)

$t=60.0\,\text{min}$ (sampling time)
$S_d=60\,\text{cm}^2$ (detector area)
$f=1.0$ (number of α-particles per decay)

The count rate, r_n, is

$$r_n = \frac{220}{600} - \frac{20}{6000} = 0.363\,\text{s}^{-1}$$

Data inserted in Eq. (3.3.17) gives

$$C = \frac{0.363 \cdot 500}{0.3 \cdot 0.8 \cdot 0.4 \cdot 1.56 \cdot 60 \cdot 60 \cdot 1.0} = 0.337\,\text{Bq/m}^3$$

The standard deviation of the net count rate r_n is given by

$$s_{r_n} = \sqrt{\frac{N_f}{t_f^2} + \frac{N_b}{t_b^2}} \tag{3.3.19}$$

Data inserted gives

$$s_{r_n} = \sqrt{\frac{220}{600^2} + \frac{20}{6000^2}} = 2.47 \cdot 10^{-2}\,\text{min}^{-1}$$

This results in a standard deviation in the concentration

$$s_C = 2.47 \cdot 10^{-2} \cdot 0.337/0.363 = 0.023\,\text{Bq m}^{-3}$$

Answer: The activity concentration of ^{239}Pu in the air is $0.34\,\text{Bq m}^{-3}$. The standard deviation is $0.02\,\text{Bq m}^{-3}$.

Solution exercise 3.6.
The half life is obtained from the relation

$$r = r_0 e^{-t\ln 2/T_{1/2}} \tag{3.3.20}$$

and thus

$$T_{1/2} = -\frac{t\ln 2}{\ln(r/r_0)} \tag{3.3.21}$$

where r_0 is the count rate at time 0 and r is the count rate at time 24 h.

Data:
$N_1=1683$ (number of total counts at time 0)
$r_b=21\,\text{min}^{-1}$ (background count rate)

$N_1 = 914$ (number of total counts at time 24 h)
$t_1 = 600\,s$ (time for measurement of total number of counts)
$t_2 = 20\,min$ (time for measurement of background counts)

Then

$$r_0 = \frac{1683}{600} - \frac{21}{60} = 2.455\,s^{-1}$$

$$r = \frac{914}{600} - \frac{21}{60} = 1.173\,s^{-1}$$

Data inserted gives

$$T_{1/2} = -\frac{24\ln 2}{\ln(1.173/2.455)} = 22.5\,h$$

Uncertainty analysis:

The total standard deviation is given by

$$s_{T_{1/2}} = \sqrt{\left(\frac{\delta T_{1/2}}{\delta r}\right)^2 s_r^2 + \left(\frac{\delta T_{1/2}}{\delta r_0}\right)^2 s_{r_0}^2} \qquad (3.3.22)$$

where

$$s_r^2 = \frac{914}{600^2} + \frac{21}{60 \cdot 20 \cdot 60} = 2.831 \cdot 10^{-3}\,s^{-2}$$

and

$$s_{r_0}^2 = \frac{1683}{600^2} + \frac{21}{60 \cdot 20 \cdot 60} = 4.967 \cdot 10^{-3}\,s^{-2}$$

Differentiation of $T_{1/2}$ gives

$$\left|\frac{\delta T_{1/2}}{\delta r}\right| = \frac{t\ln 2}{r(\ln(r/r_0))^2} = \frac{24 \cdot 3600\ln 2}{1.173(\ln(1.173/2.455))^2} = 9.360 \cdot 10^4\,s^2$$

and

$$\left|\frac{\delta T_{1/2}}{\delta r_0}\right| = \frac{t\ln 2}{r_0(\ln(r/r_0))^2} = \frac{24 \cdot 3600\ln 2}{2.455(\ln(1.173/2.455))^2} = 4.472 \cdot 10^4\,s^2$$

Data inserted in Eq. (3.3.22) gives

$$s_{T_{1/2}} = \sqrt{(9.360 \cdot 10^4)^2 \cdot 2.831 \cdot 10^{-3} + (4.472 \cdot 10^4)^2 \cdot 4.967 \cdot 10^{-3}} = 5.894 \cdot 10^3\,s$$

Answer: The half life is 22.5 h and the standard deviation is 1.6 h.

Solution exercise 3.7.

The attenuation in the material is given by

$$r_d = r_0 e^{-\mu d} \tag{3.3.23}$$

This gives

$$\mu = -\ln(r_d/r_0)/d \tag{3.3.24}$$

where μ is the linear attenuation coefficient, r_0 is the net count rate before attenuation and r_d is the net count rate after attenuation.

Data:

$d = 2.53 \pm 0.02$ cm (material thickness)
$N_1 = 35000$ (number of total counts before attenuation)
$N_2 = 25700$ (number of total counts after attenuation)
$N_b = 2350$ (number of background counts)
$t_1 = 300$ s (time for measurement of total number of counts)
$t_b = 600$ s (time for measurement of background counts)

$$r_0 = \frac{35000}{300} - \frac{2350}{600} = 112.75\ \text{s}^{-1}$$

$$r = \frac{25700}{300} - \frac{2350}{600} = 81.75\ \text{s}^{-1}$$

Data inserted in Eq. (3.3.24) gives

$$\mu = \frac{1}{2.53} \ln\left(\frac{112.75}{81.75}\right) = 0.127\ \text{cm}^{-1}$$

Uncertainty analysis:

The total uncertainty is obtained by differentiating the parameters in Eq. (3.3.24) separately.

$$\frac{\delta\mu}{\delta d} = \frac{\mu}{d} \tag{3.3.25}$$

$$\frac{\delta\mu}{\delta r_d} = \frac{1}{r_d d} \tag{3.3.26}$$

$$\frac{\delta\mu}{\delta r_0} = \frac{1}{r_0 d} \tag{3.3.27}$$

By adding the different differentials in square multiplied with their variances Eq. (3.3.28) is obtained.

$$s_\mu = \sqrt{\left(\frac{\mu}{d}\right)^2 s_d^2 + \left(\frac{1}{r_0 d}\right)^2 s_{r_0{}^2} + \left(\frac{1}{r_d d}\right)^2 s_{r_d}^2} \tag{3.3.28}$$

Data inserted gives

$$s_{r_0} = \sqrt{\frac{35000}{300^2} + \frac{2350}{600^2}} = 0.63$$

$$s_{r_d} = \sqrt{\frac{25700}{300^2} + \frac{2350}{600^2}} = 0.54$$

Data inserted in Eq. (3.3.28) gives

$$s_\mu = \sqrt{\left(\frac{0.02 \cdot 0.127}{2.53}\right)^2 + \left(\frac{0.63}{112.75 \cdot 2.53}\right)^2 + \left(\frac{0.54}{81.75 \cdot 2.53}\right)^2} = 3.56 \cdot 10^{-3}\,\text{cm}^{-1}$$

Answer: The linear attenuation coefficient is 0.127 cm^{-1} with the standard deviation 0.004 cm^{-1}.

Solution exercise 3.8.

Count rate of the detector is given by

$$r = Af\eta G e^{-\mu d} \tag{3.3.29}$$

where
A=5.0 MBq (source activity)
f=0.946·0.898 (number of 662 keV photons per decay)
d=80 mm (liquid thickness)
η=0.73 (detector efficiency)
$(\mu/\rho)_{\text{water}}$=0.00856 m^2 kg^{-1} (mass attenuation coefficient for water)
ρ_{water}=1.0·10^3 kg m^{-3} (density of water)
$(\mu/\rho)_{\text{ethanol}}$=0.00871 m^2 kg^{-1} (mass attenuation coefficient for ethanol)
ρ_{ethanol}=0.791·10^3 kg m^{-3} (density of ethanol)

The geometry factor, G, is obtained from the relation

$$G = \frac{1}{2}\left(1 - \frac{h}{\sqrt{h^2 + a^2}}\right)$$

with
h=60.0 cm (source-detector distance)
a=1.0 cm (beam radius)

$$G = \frac{1}{2}\left(1 - \frac{60}{\sqrt{60^2 + 1^2}}\right) = 6.943 \cdot 10^{-5}$$

1) Bottle filled with water

Count rate with inserted data gives

$$r_{\text{water}} = 5.0 \cdot 10^6 \cdot 0.946 \cdot 0.898 \cdot 0.73 \cdot 6.943 \cdot 10^{-5} \cdot e^{-0.00856 \cdot 1.0 \cdot 10^3 \cdot 0.080} = 108.6\,\text{s}^{-1}$$

2) <u>Bottle filled with ethanol</u>

Count rate with inserted data gives

$$r_{ethanol} = 5.0 \cdot 10^6 \cdot 0.946 \cdot 0.898 \cdot 0.73 \cdot 6.943 \cdot 10^{-5} \cdot e^{-0.00871 \cdot 0.791 \cdot 10^3 \cdot 0.080} = 124.1 \, s^{-1}$$

Background count rate

$r_b = 17 \, s^{-1}$ obtained during $t_b = 100 \, s$.

This implies that the total count rate including background is

1) Water: $r_{water,T} = 108.6 + 17 = 125.6 \, s^{-1}$

2) Ethanol: $r_{ethanol,T} = 124.1 + 17 = 141.1 \, s^{-1}$

The difference in count rate is

$r_{diff} = 141.1 - 125.6 = 15.5 \, s^{-1}$

The standard deviation in the count rate is given by

$$s_r = \sqrt{\frac{r_T}{t_T} + \frac{r_b}{t_b}} \qquad (3.3.30)$$

1) Water:

$$s_{r_{water}} = \sqrt{\frac{125.6}{t_T} + \frac{17}{100}}$$

2) Ethanol:

$$s_{r_{ethanol}} = \sqrt{\frac{141.1}{t_T} + \frac{17}{100}}$$

The standard deviation in the difference is given by

$$s_{diff} = \sqrt{s_{r_{water}}^2 + s_{r_{ethanol}}^2} \qquad (3.3.31)$$

The difference in count rate should be equal to two standard deviations of the difference. Thus

$$15.5 = 2\sqrt{\frac{125.6}{t_T} + \frac{17}{100} + \frac{141.1}{t_T} + \frac{17}{100}}$$

$t_T = 4.5 \, s$.

Answer: The measurement time needed to obtain the wanted confidence level is 4.5 s.

Solution exercise 3.9.

The net count rate at the ground surface is given by the equation

$$r_n = \frac{ASB\eta e^{-\mu d}}{4\pi d^2} \qquad (3.3.32)$$

Thus the activity is given by

$$A = \frac{4\pi d^2 r_n}{SB\eta e^{-\mu d}} \qquad (3.3.33)$$

where

A=activity (Bq)
S=10.0 cm^2 (detector area)
μ=2.83±0.05 m^{-1} (linear attenuation coefficient)
d=0.50 m (distance source-ground surface)
η=0.60±0.01 (detector efficiency)
B=1.10±0.01 (build up factor for secondary photons)

The net count rate r_n is obtained from the relation

$$r_n = \frac{N_s}{t_s} - \frac{N_b}{t_b} \qquad (3.3.34)$$

where

N_s=10 000 (number of pulses with source)
N_b=12 000 (number of background pulses)
t_s=60 s (time for measurement of source)
t_b=20.0 min (time for measurement of background)

Data inserted in Eq. (3.3.34) gives

$$r_n = \frac{10000}{60} - \frac{12000}{20.0 \cdot 60} = 156.67\,\text{s}^{-1}$$

Data inserted in Eq. (3.3.33) gives

$$A = \frac{4\pi \cdot 0.50^2 \cdot 156.67}{0.001 \cdot 1.10 \cdot 0.60 \cdot e^{-2.83 \cdot 0.5}} = 3.07\,\text{MBq}$$

To calculate the uncertainty in the determination of the activity, take the logarithm of Eq. (3.3.33) and differentiate.

$$\ln A = \ln 4\pi + 2\ln d + \ln r_n - \ln S - \ln B + \mu d - \ln \eta \qquad (3.3.35)$$

and

$$dA/A = (2/d + \mu)\,dd + dr_n/r_n - dS/S - dB/B + dd\mu - d\eta/\eta \qquad (3.3.36)$$

where the differentials can be expressed as the uncertainties.

The uncertainties can be assumed to be added in square. The standard deviation for the count rate is obtained by

$$s_{r_n} = \sqrt{\frac{10000}{60^2} + \frac{12000}{(20 \cdot 60)^2}} = 1.669\,\text{s}^{-1}$$

Inserting this value and the given uncertainties (standard deviations) in the problem, neglecting uncertainties in the detector area and the distance, gives

$$dA/A = \sqrt{(1.67/156.7)^2 + (0.01/1.10)^2 + (0.5 \cdot 0.05)^2 + (0.01/0.60)^2} = 0.0314$$

$$dA = 3.14 \cdot 10^{-2} \cdot 3.07 = 0.096\,\text{MBq}$$

Answer: The source activity is 3.1 MBq and the standard deviation 0.1 MBq.

Solution exercise 3.10.
The net count rate, r_n, at the detector is given by

$$r_n = Af\eta G \tag{3.3.37}$$

Thus the equation for the activity is

$$A = \frac{r_n}{f\eta G} \tag{3.3.38}$$

where
$f = 0.898 \cdot 0.946$ (number of 0.662 MeV photons per decay)
$\eta = 0.76$ (detector efficiency)
$\Phi = 5.08$ cm (detector diameter)
$h = 20$ or 30 cm (distance detector-source)
$G =$ geometry factor

According to the problem text the simplified relation for the geometry factor may be used. This means that G is given by ($h = 30$ cm)

$$G = \frac{\Phi^2}{4 \cdot 4 \cdot h^2} = \frac{5.08^2}{16 \cdot 30^2} = 1.792 \cdot 10^{-3} \tag{3.3.39}$$

With $h = 20$ cm, $G = 4.032 \cdot 10^{-3}$

The net count rate is
$$r_n = \frac{N_T}{t_T} - \frac{N_b}{t_b}$$

where

N_T=3562 (total number of pulses, when h=30 cm)

t_T=60.0 s (measurement time for total number of pulses)

N_b=3826 (number of background pulses)

t_b=600 s (measurement time for number of background pulses)

This gives

$$r_{30} = \frac{3562}{60} - \frac{3826}{600} = 52.99 \, s^{-1}$$

Data inserted in Eq. (3.3.38) gives the activity

$$A = \frac{52.99}{0.898 \cdot 0.946 \cdot 0.76 \cdot 1.792 \cdot 10^{-3}} = 45.80 \, kBq$$

The relative uncertainty is given by the relation

$$\left(\frac{s_A}{A}\right)^2 = \left(\frac{s_r}{r}\right)^2 + \left(\frac{s_G}{G}\right)^2 \qquad (3.3.40)$$

where s_r is

$$s_{r,30} = \sqrt{\frac{3562}{60^2} + \frac{3826}{600^2}} = 1.000 \, s^{-1}$$

and s_G is given by

$$\frac{dG}{dh} = \frac{\Phi^2 2h}{16h^4} = \frac{\Phi^2}{8h^3} \qquad (3.3.41)$$

and

$$\frac{dG}{G} = \frac{\Phi^2}{8h^3} \frac{16h^2 \, dh}{\Phi^2} = \frac{2 \, dh}{h} \qquad (3.3.42)$$

Data inserted gives

$$\left(\frac{s_A}{A}\right)^2 = \left(\frac{1.0}{52.99}\right)^2 + \left(\frac{2 \cdot 0.5}{30}\right)^2 = 1.467 \cdot 10^{-3}$$

This gives the uncertainty in the activity s_A=1.75 kBq.

With a source-detector distance of 20 cm the net count rate is obtained using the inverse square law

$$r_{20} = r_{30} \cdot 30^2/20^2 = 119.23 \, s^{-1} \qquad (3.3.43)$$

The count rate including background is then 125.60 s^{-1}. Then the standard deviation in the count rate s_r is

$$s_r = \sqrt{\frac{7536}{60^2} + \frac{3826}{600^2}} = 1.45 \, s^{-1}$$

The relative uncertainty in the activity then is

$$\left(\frac{s_A}{A}\right)^2 = \left(\frac{1.45}{119.2}\right)^2 + \left(\frac{2 \cdot 0.5}{20}\right)^2 = 2.65 \cdot 10^{-3}$$

This gives an uncertainty in the activity s_A=2.36 kBq.

Thus the uncertainty in the activity is larger with the 20 cm source-detector distance as compared with 30 cm.

Answer: The source activity is 45.8 kBq with a standard deviation of 1.7 kBq. The standard deviation is larger with a source detector distance of 20 cm as compared with 30 cm.

3.3.2 Detector Properties

Solution exercise 3.11.
The counting rate, r, will decrease with time according to the decay of the radioactive source.

$$r(t) = r_0 e^{-\lambda t} \tag{3.3.44}$$

where λ is the decay constant and r_0 is the count rate at time $t = 0$.

The average count rate during a measurement period T is given by

$$\bar{r} = \frac{\int_0^T r_0 e^{-\lambda t} dt}{T} = \frac{r_0(1 - e^{-\lambda T})}{\lambda T} \tag{3.3.45}$$

The ratio of the count rate at the center of the measuring period ($T/2$) and the average count rate is thus

$$r(T/2)/\bar{r} = \frac{\lambda T r_0 e^{-\lambda T/2}}{r_0(1 - e^{-\lambda T})} \tag{3.3.46}$$

Data:
T=30.0 min
λ=ln2/20.38 min^{-1} (decay constant for ^{11}C)

Data inserted in Eq. (3.3.46) gives

$$r(15)/\bar{r} = \frac{\ln 2 \cdot 30.0 \cdot r_0 \cdot e^{-15 \cdot \ln 2/20.38}}{20.38 \cdot r_0(1 - e^{-30 \cdot \ln 2/20.38})} = 0.958$$

Answer: The ratio between the count rate at the center of the measurement period and the average count rate is 0.96.

Solution exercise 3.12.
The total count rate is given by

$$r_{\text{tot}} = r_{12} + r_{\text{ch}} + r_b \tag{3.3.47}$$

where r_{12} is the true coincidence rate given by

$$r_{12} = A\eta^2 G^2 f_1 f_2 \qquad (3.3.48)$$

and r_{ch} is the random coincidence rate given by

$$r_{ch} = 2r_1 r_2 \tau = 2A^2 \eta^2 G^2 f_1 f_2 \tau \qquad (3.3.49)$$

r_1, r_2 are count rates of the separate detectors

Data:
A=3.90 MBq (source activity)
η=0.45 (detector efficiency)
f_1, f_2=1.0 (number of photons per decay)
τ=2.0 μs (resolving time of the coincidence circuit)
r_b=2.0 s^{-1} (coincidence background count rate)

G is the geometry factor given by

$$G = \frac{1}{2}(1 - \frac{h}{\sqrt{h^2 + a^2}}) \qquad (3.3.50)$$

where
h=20 cm (distance detector surface-source)
a=2.5 cm (detector radius)

Data inserted in Eq. (3.3.50) gives

$$G = \frac{1}{2}(1 - \frac{20}{\sqrt{20^2 + 2.5^2}}) = 3.86 \cdot 10^{-3} \qquad (3.3.51)$$

Data inserted in Eq. (3.3.47) gives

$r = 3.90{\cdot}10^6{\cdot}0.45^2{\cdot}(3.86{\cdot}10^{-3})^2{\cdot}1.0^2 + 2{\cdot}(3.90{\cdot}10^6{\cdot}0.45{\cdot}3.86{\cdot}10^{-3})^2{\cdot}1{\cdot}1{\cdot}2.0{\cdot}10^{-6} + 2.0$

r=197 s^{-1}

Answer: The total coincidence rate is $0.20{\cdot}10^3$ s^{-1}.

Solution exercise 3.13.
The true coincidence rate in the sum peak is given by

$$r_{12} = A\eta_1 \eta_2 f_1 f_2 G^2 \qquad (3.3.52)$$

The random coincidence rate is given by

$$r_{ch} = A^2 \eta_1 \eta_2 f_1 f_2 G^2 \tau_r \qquad (3.3.53)$$

The total count rate r_T is given by

$$r_T = r_{12} + r_{ch} + r_b \tag{3.3.54}$$

Data:
A=source activity
η_1=0.30 (detector efficiency for photon 1)
η_2=0.50 (detector efficiency for photon 2)
f_1, f_2=1.0 (number of photons per decay)
τ_r=2.0 μs (resolving time of the coincidence circuit)
r_b=12.0 s^{-1} (coincidence background count rate)
r_T=129 s^{-1} (total coincidence count rate)
G is the geometry factor given by

$$G = \frac{1}{2}(1 - \frac{h}{\sqrt{h^2 + a^2}}) \tag{3.3.55}$$

where
h=10 cm (distance detector surface-source)
a=5.0 cm (detector radius)

Data inserted in Eq. (3.3.55)gives

$$G = \frac{1}{2}(1 - \frac{10}{\sqrt{10^2 + 5^2}}) = 5.28 \cdot 10^{-2}$$

Data inserted in Eq. (3.3.54) gives

$$129 = A \cdot 0.3 \cdot 0.5 \cdot 1.0 \cdot 1.0 \cdot (5.28 \cdot 10^{-2})^2 + A^2 \cdot 0.3 \cdot 0.5 \cdot 1.0 \cdot 1.0 \cdot (5.28 \cdot 10^{-2})^2 \cdot 2.0 \cdot 10^{-6} + 12$$

Reorganizing the equation gives

$$A^2 + \frac{A \cdot 4.180 \cdot 10^{-4}}{8.359 \cdot 10^{-10}} = \frac{117}{8.359 \cdot 10^{-10}}$$

and

$$A = -\frac{4.180 \cdot 10^{-4}}{2 \cdot 8.359 \cdot 10^{-10}} \pm \sqrt{\left(\frac{4.180 \cdot 10^{-4}}{2 \cdot 8.359 \cdot 10^{-10}}\right)^2 + \frac{117}{8.359 \cdot 10^{-10}}}$$

$$A = -2.500 \cdot 10^5 \pm 4.500 \cdot 10^5$$

A=2.000·10^5 Bq

Answer: The source activity is 200 kBq.

Solution exercise 3.14.

The true count rate, neglecting background is given by

$$r_0 = A\eta ftG \qquad (3.3.56)$$

where
A=3.0 kBq (source activity)
f=1.0 (number of β-particles per decay)
η=1.0 (detector efficiency)
t=0.87 (transmission through detector window)
h=40 mm (distance detector surface-source)
a=15.0 mm (detector radius)
and
G is the geometry factor given by

$$G = \frac{1}{2}(1 - \frac{h}{\sqrt{h^2 + a^2}}) \qquad (3.3.57)$$

Data inserted gives

$$G = \frac{1}{2}(1 - \frac{40}{\sqrt{40^2 + 15^2}}) = 3.18 \cdot 10^{-2}$$

Data inserted in Eq. (3.3.56) gives

$$r_0 = 3.0 \cdot 10^3 \cdot 1.0 \cdot 1.0 \cdot 0.87 \cdot 3.18 \cdot 10^{-2} = 83.0\,s^{-1}$$

The relation between measured and true count rate is given by

$$r_0 = \frac{r}{1 - r\tau} \qquad (3.3.58)$$

According to the problem $r = 0.99r_0$. Thus

$$r_0 = \frac{0.99r_0}{1 - 0.99r_0\tau}$$

and

$$\tau = (1 - 0.99)/(0.99 \cdot 83.0) = 1.22 \cdot 10^{-4}\ s$$

Answer: The dead time is 0.12 ms.

Solution exercise 3.15.

The field strength in the cylindrical counter at radius r is given by (see Fig. 3.3)

$$\epsilon(r) = \frac{V}{r \ln(r_2/r_1)} \qquad (3.3.59)$$

The ionization energy is then given by

$$dE = q \frac{V}{r \ln(r_2/r_1)} \, dr \tag{3.3.60}$$

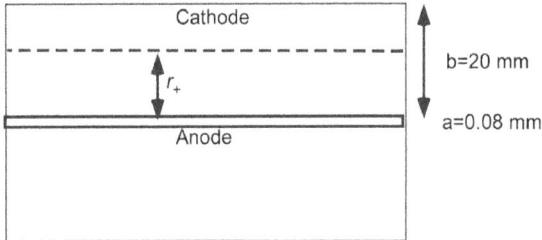

Figure 3.3: Sketch of the GM-counter in exercise 3.15.

dE is the energy transferred to the electron when passing a distance dr. The electron has to obtain an energy of $W=15.7$ eV in a distance, l, equal to the free mean path.

Integration of Eq. (3.3.60) gives

$$E = \int_r^{r+l} \frac{qV \, dr}{r \ln(r_2/r_1)} = \frac{qV}{\ln(r_2/r_1)} \ln(\frac{r+l}{r}) \tag{3.3.61}$$

Solving V gives

$$V = \frac{E \ln(r_2/r_1)}{q \ln((r+l)/r)} \tag{3.3.62}$$

Data:
$r=0.65$ mm (the maximal radius where secondary electrons can be produced)
$r_1=0.08$ mm (anode radius)
$r_2=20$ mm (cathode radius)
$V= 1500$ V (voltage over the counter)
$q=1$ (charge per electron if energy W is measured in eV)
$l=6.4 \cdot 10^{-4} \cdot 101.3/p$ mm, where p is the gas pressure in the counter.

Data inserted in Eq. (3.3.62) gives

$$1500 = \frac{15.7 \ln(20.0/0.08)}{\ln(\frac{0.65+\frac{6.4 \cdot 10^{-4} \cdot 101.3}{p}}{0.65})}$$

Rearranging the equation gives

$$\frac{15.7 \ln(20.0/0.08)}{1500} = \ln(\frac{0.65 + \frac{6.4 \cdot 10^{-4} \cdot 101.3}{p}}{0.65})$$

and

$$0.65 \cdot e^{\frac{15.7 \ln(20.0/0.08)}{1500}} = 0.65 + \frac{6.4 \cdot 10^{-4} \cdot 101.3}{p}$$

Solving p gives

$p = 1.68$ kPa

Answer The gas pressure in the counter is 1.7 kPa.

Solution exercise 3.16.
The field strength in the cylindrical counter at radius r is given by

$$\epsilon(r) = \frac{V}{r \ln(r_2/r_1)} \tag{3.3.63}$$

The electron shall in the path length 1.0 mm obtain the energy $E=23$ eV, where E is given by

$$E = q \int_r^{r+1} E(r)\, dr = q \int_r^{r+1} \frac{V}{r \ln(r_2/r_1)}\, dr = \frac{qV}{\ln(r_2/r_1)} (\ln(r+1) - \ln r) \tag{3.3.64}$$

Data:
$r_1 = 25\,\mu$m (anode radius)
$r_2 = 25$ mm (cathode radius)
$V = 1000$ V (voltage over the counter)
$q = 1$ (charge per electron if energy E is measured in eV)

Data inserted in Eq. (3.3.64) gives

$$23 = \frac{1 \cdot 1000}{\ln(25 \cdot 10^{-3}/25 \cdot 10^{-6})} (\ln(r+1) - \ln r) \tag{3.3.65}$$

$$0.15888 = \ln(\frac{r+1}{r}) \tag{3.3.66}$$

$r = 5.81$ mm

Answer: The radius where the electron gets enough energy is 5.8 mm.

Solution exercise 3.17.
The pulse height from a GM-counter at a time t after the pulse is initiated is given by the relation

$$V_r(t) = \frac{Q}{C} \frac{1}{\ln(r_2/r_1)} \ln \left(\frac{2\mu V_0 t}{r_1^2 p \ln(r_2/r_1)} + 1 \right)^{1/2} \tag{3.3.67}$$

where
r_1=20 µm (anode radius)
r_2=20 mm (cathode radius)
V_0= 600 V (voltage over the counter)
p=10.7 kPa (gas pressure)
μ=1.40·10^4 (mm s^{-1})(V mm^{-1})$^{-1}$kPa (ion mobility)
Q=produced charge
C=detector capacitance

Data inserted in Eq. (3.3.67) gives

$$V_r(t) = \frac{Q}{C} \frac{1}{\ln(20/20 \cdot 10^{-3})} \ln \left(\frac{2 \cdot 1.40 \cdot 10^4 \cdot 600 \cdot t}{(20 \cdot 10^{-3})^2 \cdot 10.7 \ln(20/20 \cdot 10^{-3})} + 1 \right)^{1/2}$$

$V_r(t)$=0.25(Q/C) according to information.

Inserting this in the equation and solve for t gives

t=5.39·10^{-8} s.

The radius where the ions are positioned at a certain time t is given by the relation

$$r(t) = \sqrt{\frac{2\mu}{p} \frac{V_0 t}{\ln(r_2/r_1)} + r_1^2} \qquad (3.3.68)$$

Data inserted in Eq. (3.3.68) gives

$$r(t) = \sqrt{\frac{2 \cdot 1.40 \cdot 10^4}{10.7} \frac{600 \cdot 5.39 \cdot 10^{-8}}{\ln(20/20 \cdot 10^{-3})} + (20 \cdot 10^{-3})^2} = 0.11 \, \text{mm}$$

The anode radius is 20 µm. This means that the ions have traveled a distance of

d=0.11-20·10^{-3}=0.09 mm

Answer: The ions have traveled a distance of 0.09 mm.

Solution exercise 3.18.
The current obtained when the detector is used as an ionization chamber is given by

$$I = \frac{\sum Ay_i \bar{E}_{\beta_i} Gq}{\bar{W}} \qquad (3.3.69)$$

where
A=2.76 MBq (source activity)

$\bar{W}$=35 eV (mean energy to produce an ion pair in argon)
q=1.602·10^{-19} C (charge of the electron)
y_i=number of β-particles per decay with the mean energy $\bar{E}_{\beta_i}$. See Table 3.1.
G is the geometry factor given by

$$G = \frac{1}{2}\left(1 - \frac{h}{\sqrt{h^2 + a^2}}\right) \tag{3.3.70}$$

Data inserted for the geometry (h=50 mm, a=20 mm) gives the value of G

$$G = \frac{1}{2}\left(1 - \frac{5.0}{\sqrt{5.0^2 + 2.0^2}}\right) = 0.0358$$

The data for the decay of ^{59}Fe gives the total β-energy E_β=0.117 MeV.

Table 3.1: Mean energy and frequency of β-particles from ^{59}Fe decay.

E_{β_i} (Energy/MeV)	y_i(frequency)	$y_i \cdot E_{\beta_i}$
0.036	0.0127	4.54·10−4
0.081	0.456	3.69·10^{-2}
0.149	0.528	7.88·10^{-2}
0.636	0.0018	1.14·10^{-3}
Total		0.117

Data inserted in Eq. (3.3.69) gives

$$I = \frac{2.76 \cdot 10^6 \cdot 0.117 \cdot 10^6 \cdot 0.0358 \cdot 1.602 \cdot 10^{-19}}{35.0} = 5.26 \cdot 10^{-11} \text{ A}$$

The number of pulses per second when the detector is used as a GM-counter is obtained from the relation

$$r_0 = A\sum y_i G\eta \tag{3.3.71}$$

where η=1.0 is the detector efficiency

Data inserted gives

$$r_0 = 2.76 \cdot 10^6 \cdot 0.0358 \cdot 0.9985 \cdot 1.0 = 9.866 \cdot 10^4 \text{ s}^{-1}$$

This is the true count rate. The measured count rate is less due to dead time losses. The relation between measured and true count rate is given by

$$r = \frac{r_0}{1 + r_0\tau} \tag{3.3.72}$$

With the dead time τ=130 μs the measured count rate, neglecting background count rate, is

$$r = \frac{9.866 \cdot 10^4}{1 + 9.866 \cdot 10^4 \cdot 130 \cdot 10^{-6}} = 7.136 \cdot 10^3 \, s^{-1}$$

The relation between the ionization energy (W) and the voltage in the detector is given by the relation

$$W = \frac{eV}{\ln(r_2/r_1)}(\ln x_2 - \ln x_1) \tag{3.3.73}$$

where
W=15.7 eV
r_1=0.08 cm (anode radius)
r_2=2.0 cm (cathode radius)
$x_2 - x_1$=6.4·10^{-4} mm (mean free path at 101.33 kPa)

At a pressure of 1.5 kPa, $x_2 - x_1$=0.0432 mm. With x_2=0.65 mm according to the problem, x_1=0.65-0.0432=0.6068 mm

Data inserted in Eq. (3.3.73) gives

$$15.7 = \frac{1.0 \cdot V}{\ln(2.0/0.08)}(\ln 0.65 - \ln 0.6068)$$

V=735 V.

Answer The ionization current is 5.3·10^{-11} A. The measured count rate is 7.14·10^3 s^{-1}. The voltage over the counter is 735 V.

Solution exercise 3.19.
The measured voltage is obtained from the relation

$$V = \frac{Q}{C} = \frac{Eq}{\bar{W}C} \tag{3.3.74}$$

where
E=0.70 MeV (electron energy)
q=1.602·10^{-19} C (charge of an electron)
C=13.0 pF (capacitance over the chamber)
$\bar{W}$=33.97 eV (mean energy to produce an ion pair in air)

Data inserted in Eq. (3.3.74) then gives

$$V = \frac{0.70 \cdot 10^6 \cdot 1.602 \cdot 10^{-19}}{33.97 \cdot 13.0 \cdot 10^{-12}} = 2.54 \cdot 10^{-4} \, V$$

The time to collect the ions is given by the distance traveled and the velocity. The velocity is given by the relation

$$v = \frac{\mu V}{dp} \tag{3.3.75}$$

where
$\mu_i = 1.32 \cdot 10^4$ (mm s^{-1})(V mm^{-1})$^{-1}$kPa (ion mobility)
$\mu_e = 1.62 \cdot 10^7$ (mm s^{-1})(V mm^{-1})$^{-1}$kPa (electron mobility)
V=400 V (voltage over the chamber)
p=100 kPa (gas pressure in the chamber)
d=2.0 mm (distance between electrodes)
The time needed to cross the chamber is given by

$$t = \frac{d}{v} = \frac{d^2 p}{\mu V} \tag{3.3.76}$$

The shortest time is obtained when the ions are produced close to the negative electrode resulting in transport only by electrons. The longest time is then obtained when the ions are produced close to the positive electrode and the transport is only by ions.

a) Shortest time

$$t = \frac{2^2 \cdot 100}{1.62 \cdot 10^7 \cdot 400} = 6.17 \cdot 10^{-8}\,\text{s}$$

b) Longest time

$$t = \frac{2^2 \cdot 100}{1.32 \cdot 10^4 \cdot 400} = 7.58 \cdot 10^{-5}\,\text{s}$$

Answer The shortest collection time is 62 ns and the longest 76 μs. The pulse voltage is 0.25 mV

Solution exercise 3.20.
The pulse height at a time t is given by the relation

$$V(t) = \frac{1}{\Lambda - \theta} \frac{\Lambda Q}{C} (e^{-\theta t} - e^{-\Lambda t}) \tag{3.3.77}$$

and the time t_{max} when the maximum voltage is obtained, is given by solving the equation

$$\frac{dV}{dt} = \frac{1}{\Lambda - \theta} \frac{\Lambda Q}{C} (-\theta e^{-\theta t} + \Lambda e^{-\Lambda t}) = 0 \tag{3.3.78}$$

Thus

$$\theta e^{-\theta t} - \Lambda e^{-\Lambda t} = 0 \tag{3.3.79}$$

and

$$t_{max} = \frac{\ln \frac{\theta}{\Lambda}}{\theta - \Lambda} \tag{3.3.80}$$

where Λ is the decay constant of the light from the NaI-scintillator and θ the time constant of the anode circuit.

The charge Q is given by the relation

$$Q = \frac{E_0 \eta \epsilon p \alpha \delta^N q}{E_\lambda} \tag{3.3.81}$$

where
E_0=1.33 MeV (energy of the photon)
η=0.12 (light conversion efficiency)
ϵ=0.50 (light collection efficiency)
p=0.20 (photo cathode efficiency)
α=0.80 (collection efficiency at the first dynode)
δ=2.5 (amplification at each dynode)
N=10 (number of dynodes)
$q = 1.602 \cdot 10^{-19}$ C (charge of an electron)

E_λ is obtained from the knowledge of the wavelength of the light photons. The relation between energy and wavelength is

$$E_\lambda = \frac{hc}{\lambda} \tag{3.3.82}$$

where
$h = 6.6256 \cdot 10^{-34}$ Js (Planck's constant)
$c = 3.0 \cdot 10^8 \, \mathrm{m \, s^{-1}}$ (velocity of light)
λ=415 nm (wavelength of the light photons)

Data inserted in Eq. (3.3.82) gives

$$E_\lambda = \frac{6.6256 \cdot 10^{-34} \cdot 3.0 \cdot 10^8}{415 \cdot 10^{-9} \cdot 1.602 \cdot 10^{-19}} = 2.99 \, \mathrm{eV}$$

Data inserted in Eq. (3.3.81) gives

$$Q = \frac{1.33 \cdot 10^6 \cdot 0.12 \cdot 0.50 \cdot 0.20 \cdot 0.8 \cdot 2.5^{10} \cdot 1.602 \cdot 10^{-19}}{2.99} = 6.5245 \cdot 10^{-12} \, \mathrm{C}$$

θ in Eq. (3.3.77) is given by θ =1/RC, where RC is the time constant of the anode circuit. With R=100 kΩ and C=100 pF, θ=1.0$\cdot$10^5 s^{-1}.

$\Lambda = 4.35 \cdot 10^6 \, \mathrm{s^{-1}}$ (decay constant of the NaI-scintillator)

Inserting θ and Λ in Eq. (3.3.80) gives

$$t_{max} = \frac{\ln(\frac{1.0 \cdot 10^5}{4.35 \cdot 10^6})}{1.0 \cdot 10^5 - 4.35 \cdot 10^6} = 0.888 \cdot 10^{-6} \text{ s}$$

t_{max} inserted in Eq. (3.3.77) gives

$$V(t_{max}) = \frac{1}{(4.35 \cdot 10^6 - 1.0 \cdot 10^5)} \frac{4.35 \cdot 10^6 \cdot 6.524 \cdot 10^{-12}}{100 \cdot 10^{-12}}$$
$$\times (e^{-0.888 \cdot 10^{-6} \cdot 10^5} - e^{-0.888 \cdot 10^{-6} \cdot 4.35 \cdot 10^6})$$

$$V(t_{max}) = 5.97 \cdot 10^{-2} \text{ V}$$

The voltage corresponding to a channel in the multichannel analyzer is obtained from the knowledge of the voltage at the maximum channel number. Thus the voltage at channel 500, when 10.0 V corresponds to channel 1024 is 500·10.0/1024=4.88 V. The voltage from the PM-tube is 5.97·10^{-2} V. The needed amplification G is then

G=4.88/5.97·10^{-2}=81.8

Answer: The amplification is 82.

Solution exercise 3.21.
The number of produced electron-hole pairs in the detector is given by

$$N = \frac{E}{\epsilon} \tag{3.3.83}$$

The voltage after amplification is given by

$$V = \frac{EGq}{\epsilon C} \tag{3.3.84}$$

where
E=the absorbed energy in the HpGe-detector in eV.
ϵ=2.98 eV (energy per electron-hole pair)
G=1500 (amplification in the amplifier)
$C = 7.5 \cdot 10^{-12}$ F (input capacitance)
$q = 1.602 \cdot 10^{-19}$ C (charge of an electron)
V=1.25 V

Solving Eq. (3.3.84) for E and inserting data gives

$$E = \frac{1.25 \cdot 7.5 \cdot 10^{-12} \cdot 2.98}{1500 \cdot 1.602 \cdot 10^{-19}} = 1.165 \cdot 10^5 \text{ eV} \tag{3.3.85}$$

The energy of the primary photon is obtained from the Compton relation

$$hv' = \frac{hv_0}{1 + \frac{hv_0}{m_ec^2}(1 - \cos\theta)} \tag{3.3.86}$$

The photon is scattered in $\theta = 90^0$ and thus $\cos\theta = 0$. Inserting $hv' = 0.1165$ MeV gives

$$0.1165 = \frac{hv_0}{1 + \frac{hv_0}{0.511}}$$

$hv_0 = 0.1505$ MeV

The FWHM is given by the relation $w_{1/2} = 2.35\sqrt{EF\epsilon}$, where F=0.10 (Fano factor)

Thus

$w_{1/2} = 2.35\sqrt{0.1165 \cdot 10^6 \cdot 0.1 \cdot 2.98} = 0.44$ keV

Answer: The primary photon energy is 0.15 MeV and the FWHM is 0.44 keV.

Solution exercise 3.22.
The absorbed energy in a layer dx at depth x is given by

$$dE = \sum(\mu_{en}/\rho)_i \Psi_{0,i} S\rho e^{-\mu_i x}\, dx \tag{3.3.87}$$

Integration over the whole detector thickness d gives

$$E = \int_0^d \sum(\mu_{en}/\rho)_i \Psi_{0,i} S\rho e^{-\mu_i x}\, dx = \sum \frac{(\mu_{en}/\rho)_i \Psi_{0,i} S\rho}{\mu}(1 - e^{-\mu_i d}) \tag{3.3.88}$$

where
μ_{en}/ρ = mass energy absorption coefficient for NaI
μ/ρ = mass attenuation coefficient in NaI
$\Psi_{0,i} = AE_iy_i/(4\pi l^2)$ (impinging energy fluence for energy E_i)
A = activity
l = 200 mm (distance source detector surface)
d = 50 mm (detector thickness)
$S = \pi r^2 = \pi \cdot 0.01^2$ m^2 (irradiated detector area)

^{65}Zn decays and emits photons of two energies, which can contribute to the energy deposition. See Table 3.2 which also includes data for the interaction coefficients.

Table 3.2: Decay data and interaction coefficients in NaI for photons from ^{65}Zn.

Energy (MeV)	y (frequency)	$(\mu/\rho)_{\text{NaI}}$(m^2 kg^{-1})	$(\mu_{\text{en}}/\rho)_{\text{NaI}}$(m^2 kg^{-1})
0.511	0.029	0.00935	0.004085
1.116	0.507	0.00560	0.00257

Calculation of absorbed energy.

a) E=0.511 MeV

$$E_{0.511} = \frac{0.004085 \cdot 0.029 \cdot 0.511 \cdot \pi \cdot 0.01^2 \cdot A}{4 \cdot \pi \cdot 0.2^2 \cdot 0.00935}(1 - e^{-0.00935 \cdot 0.05 \cdot 3.67 \cdot 10^3})$$

$E_{0.511}=3.38 \cdot 10^{-6} A$ MeV

b) E=1.116 MeV

$$E_{1.116} = \frac{0.00257 \cdot 0.507 \cdot 1.116 \cdot \pi \cdot 0.01^2 \cdot A}{4 \cdot \pi \cdot 0.2^2 \cdot 0.0056}(1 - e^{-0.00560 \cdot 0.05 \cdot 3.67 \cdot 10^3})$$

$E_{1.116}=1.042 \cdot 10^{-4} A$ MeV

The total deposited energy is $E_{0.511} + E_{1.116}=1.076 \cdot 10^{-4} A$ MeV

The energy to produce a photoelectron at the photocathode is ϵ=100 eV. Thus the number of photoelectrons produced is

$$N = \frac{E}{\epsilon} = \frac{1.076 \cdot 10^{-4} \cdot 10^6 \cdot A}{100} = 1.076A$$

The current at the output of the photomultiplier is given by

$$I = NqM \tag{3.3.89}$$

where
q=1.602 $\cdot 10^{-19}$ C (charge of an electron)
$M = \delta^L = 3.5^{10}$ (δ is the electron amplification per dynode and L is the number of dynodes)

Data inserted in Eq (3.3.89) gives

$$I = 1.076 \cdot 1.602 \cdot 10^{-19} \cdot 3.5^{10} \cdot A = 4.755 \cdot 10^{-14} \cdot A$$

This is the current induced by the radiation. There is also a background current of 124 μA. The standard deviation in the background current is given by

$$\frac{\sigma_r}{r} = \frac{1}{\sqrt{2rRC}} \tag{3.3.90}$$

where r is the pulse rate, and $RC = 20.0$ s is the time constant.

r is obtained from the knowledge of the current $I_b = 124$ μA and the charge per pulse $Q = 2.39 \cdot 10^{-11}$ C.

This gives

$r = 124 \cdot 10^{-6}/2.39 \cdot 10^{11} = 5.188 \cdot 10^6 \, s^{-1}$.

Data inserted in Eq (3.3.90) gives

$$\sigma_r = \frac{5.188 \cdot 10^6}{\sqrt{2 \cdot 5.188 \cdot 10^6 \cdot 20}} = 360.1 \, s^{-1}$$

This corresponds to a standard deviation in the current of

$\sigma_I = 360.1 \cdot 2.39 \cdot 10^{-11} = 8.608 \cdot 10^{-9}$ A.

The net current should be equal to twice the standard deviation of the background current, i. e. $17.22 \cdot 10^{-9}$ A.

The activity can now be obtained from the relation

$I = 4.755 \cdot 10^{-14} \cdot A = 17.22 \cdot 10^{-9}$

$A = 0.362 \cdot 10^6$ Bq

Answer: The activity is 0.36 MBq.

Solution exercise 3.23.
The activity of a sample in an activation process is given by

$$A = \sigma \dot{\Phi} N (1 - e^{-\lambda t}) \tag{3.3.91}$$

and

$$A_\infty = \sigma \dot{\Phi} N = \sigma \dot{\Phi} \frac{m N_A}{m_a} \tag{3.3.92}$$

The saturation activity A_∞, may be obtained from Eq (3.3.93) by measuring the irradiated sample (see Fig. 3.4).

$$A_\infty = \frac{\lambda(C - B)}{y \eta (1 - e^{-\lambda t_0})(e^{-\lambda t_1} - e^{-\lambda t_2})} \tag{3.3.93}$$

Data:

$\sigma = 2640 \cdot 10^{-28}$ m^2 (activation cross section)

$C = 76800$ (total number of counts)

$B = 13 \cdot 20 \cdot 60$ (number of background counts)

$t_0 = 10.0$ min, $t_1 = 10.0$ min, $t_2 = 30.0$ min (See Fig. 3.4)

$\eta = 0.30$ (detector efficiency)

$y = 0.48$ (number of photons/disintegration)

$\lambda = \ln 2/54.1 = 0.0128$ min^{-1} (decay constant for ^{116m}In)

$m = 2.0$ g (sample mass)

$N_A = 6.022 \cdot 10^{23}$ mol^{-1} (Avogadro's number)

$m_a = 115$ (atomic mass of ^{115}In)

Data inserted in Eq. (3.3.93) gives

$$A_\infty = \frac{0.0128(76800 - 15600)}{60 \cdot 0.48 \cdot 0.30(1 - e^{-0.0128 \cdot 10})(e^{-0.0128 \cdot 10} - e^{-0.0128 \cdot 30})}$$

$A_\infty = 3797$ Bq

Inserting this in Eq. (3.3.92) and solving for $\dot{\Phi}$ gives

$$\dot{\Phi} = \frac{3797 \cdot 115}{2640 \cdot 10^{-28} \cdot 2.0 \cdot 6.022 \cdot 10^{23}} = 1.373 \cdot 10^6 \, \text{m}^{-2} \, \text{s}^{-1}$$

Answer: The fluence rate is $1.37 \cdot 10^6$ m^{-2}s^{-1}

Solution exercise 3.24.

The saturation activity A_∞, is given from Eq (3.3.94) and Eq (3.3.95).

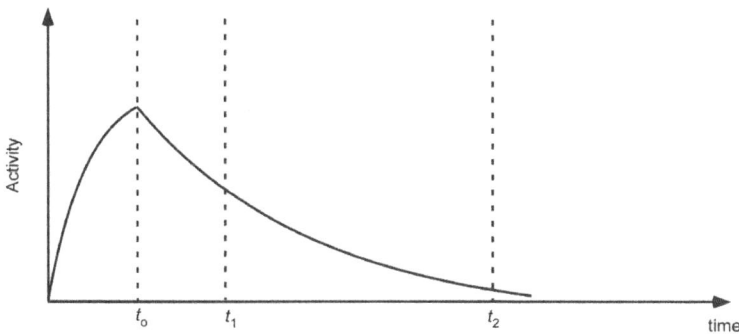

Figure 3.4: Variation of activity with time during and after activation.

$$A_\infty = \frac{\lambda(C - B)}{y\eta(1 - e^{-\lambda t_0})(e^{-\lambda t_1} - e^{-\lambda t_2})} \tag{3.3.94}$$

and

$$A_\infty = \sigma\dot{\Phi}N = \sigma\dot{\Phi}\frac{mN_A}{m_a} \tag{3.3.95}$$

Two isotopes of silver are produced during the activation process, ^{108}Ag(1) and ^{110}Ag(2). The net count obtained C-B is then given by

$$C - B = \dot{\Phi}mN_A\eta\left[\frac{f_1\sigma_1y_1(1 - e^{-\lambda_1 t_0})(e^{-\lambda_1 t_1} - e^{-\lambda_1 t_2})}{\lambda_1 m_{a,1}}\right.$$
$$\left. + \frac{f_2\sigma_2y_2(1 - e^{-\lambda_2 t_0})(e^{-\lambda_2 t_1} - e^{-\lambda_2 t_2})}{\lambda_2 m_{a,2}}\right] \tag{3.3.96}$$

Data:

$C=25000$ (total amount of counts)

$B=3\cdot300=900$ (background counts)

$N_A=6.022\cdot10^{23}$ mol^{-1} (Avogadro's number)

$\eta=0.10$ (detector efficiency)

$\dot{\Phi} = 10.2 \cdot 10^7$ m^{-2} s^{-1} (neutron fluence rate)

$f_1=0.5135$ (mass fraction of ^{107}Ag)

$f_2=0.4865$ (mass fraction of ^{109}Ag)

$\lambda_1 = \ln 2/(2.37 \cdot 60) = 4.874 \cdot 10^{-3}$ s^{-1} (decay constant for ^{108}Ag)

$\lambda_2 = \ln 2/(24.6) = 2.818 \cdot 10^{-2}$ s^{-1} (decay constant for ^{110}Ag)

$m_{a,1}=107$ (atomic mass for ^{107}Ag)

$m_{a,2}=109$ (atomic mass for ^{109}Ag)

$y_1=0.977$ (frequency of emitted electrons/decay of ^{108}Ag)

$y_2=0.997$ (frequency of emitted electrons/decay of ^{110}Ag)

$\sigma_1 = 3.0 \cdot 10^{-28}$ m^2 (activation cross section for ^{107}Ag)

$\sigma_2 = 110.0 \cdot 10^{-28}$ m^2 (activation cross section for ^{109}Ag)

Data inserted in Eq. (3.3.96) gives m.

$$25000-900 = 6.022 \cdot 10^{23} \cdot 10.2 \cdot 10^7 \cdot 0.1 \cdot m \cdot 10^{-28}$$

$$\left[\frac{0.5135 \cdot 30 \cdot 0.977(1 - e^{-600\cdot4,874\cdot10^{-3}})(e^{-60\cdot4.874\cdot10^{-3}} - e^{-360\cdot4.874\cdot10^{-3}})}{4.874 \cdot 10^{-3} \cdot 107}\right.$$
$$\left. + \frac{0.4865 \cdot 110 \cdot 0.997(1 - e^{-600\cdot2.818\cdot10^{-2}})(e^{-60\cdot2.818\cdot10^{-2}} - e^{-360\cdot2.818\cdot10^{-2}})}{2.818 \cdot 10^{-2} \cdot 109}\right]$$

$m=2.08$ g

Answer The silver mass in the coin is 2.1 g.

Solution exercise 3.25.

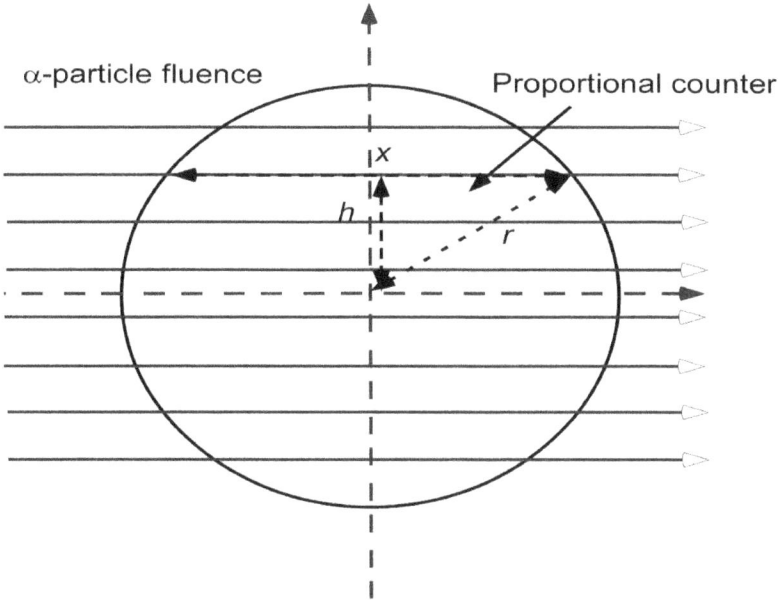

Figure 3.5: Detector geometry with the "wall less" proportional counter irradiated with a parallel fluence of α-particles.

The relative number of particles that hits the spherical counter between h and $h + dh$ (see Fig. 3.5) may be expressed as

$$P(h)\,dh = \frac{2\pi\,dh\,\Phi}{\pi r^2\,\Phi} = \frac{2h\,dh}{r^2} \tag{3.3.97}$$

where r is the radius of the proportional counter, and Φ is the incoming particle fluence.

The track length, x, of these particles in the spherical counter is, assuming straight particle tracks, given by

$$\left(\frac{x}{2}\right)^2 = r^2 - h^2 \tag{3.3.98}$$

Differentiation of the relation gives

$$\frac{2x\,dx}{4} = 2h\,dh \tag{3.3.99}$$

The number of particles between h and $h + dh$ is equal to the number of particles with the track length between x and $x + dx$, i e $P(h)\,dh = P(x)\,dx$. Thus

$$P(x)\,dx = \frac{2h\,dh}{r^2} = \frac{x\,dx}{2r^2} \tag{3.3.100}$$

If it is assumed that the energy loss of the particles is constant, the probability distribution of pulse heights (V) will be equal to the probability distribution of track lengths, i e $P(V) \approx P(x)$, which increases linearly with x or V.

Energy loss for an α-particle passing the full diameter, d, of the counter is given by

$$E = d \cdot \rho \frac{1}{\rho} \frac{dE}{dx} \tag{3.3.101}$$

Data:

$d = 10 \cdot 10^{-3}$ m (diameter of the sphere)

$\rho = 1.128 \frac{273}{293} \frac{9.44}{101.3} = 9.794 \cdot 10^{-2}$ kg m^{-3} (density of the tissue equivalent gas)

$\frac{1}{\rho} \frac{dE}{dl} = 88.33$ MeVm^2 kg^{-1} (mass stopping power of the tissue equivalent gas)

Data inserted in Eq (3.3.101) gives

$$E = 10 \cdot 10^{-3} \cdot 9.794 \cdot 10^{-2} \cdot 88.33 = 0.0865\,\text{MeV}$$

The obtained pulse height is given by

$$V = \frac{EqMG}{\bar{W}C} \tag{3.3.102}$$

where

$\bar{W} = 31.1$ eV (mean energy per ion pair)

$q = 1.602 \cdot 10^{-19}$ C (charge of an electron)

$M = 1000$ (gas amplification)

$C = 1.0 \cdot 10^{-12}$ F (capacitance of the preamplifier)

$G = 10$ (amplification in the preamplifier)

Data inserted in Eq (3.3.102) gives

$$V = \frac{0.0865 \cdot 10^6 \cdot 1.602 \cdot 10^{-19} \cdot 1000 \cdot 10}{31.1 \cdot 1.0 \cdot 10^{-12}} = 4.456\,\text{V}$$

Channel 1024 corresponds to a voltage of 10 V. Then 4.456 V corresponds to channel 4.456·1024/10=456.

Answer: The distribution is triangular with a maximum at channel 456.

4 Radiation Dosimetry

4.1 Definitions and Relations

4.1.1 Definitions of Important Dosimetric Quantities

When ionizing radiation interacts with matter it deposits energy. Uncharged radiation like photons and neutrons perform this mainly in two steps where first charged particles are produced. Charged particles then deposit energy through excitations and ionizations. An important task in dosimetry is to determine the deposited energy both at the macroscopic and microscopic level. An accurate determination of deposited or absorbed energy is important both in radiotherapy and in radiation protection.

Some important quantities in dosimetry and some relations that will be used in the exercises are described below. The material is mainly based on ICRU Report 85: Fundamental Quantities and Units for Ionizing Radiation (ICRU, 2011) and Attix: Introduction to Radiological Physics and Radiation Dosimetry (Attix, 1986).

Energy deposit
The stochastic quantity *energy deposit* is the basic quantity from which all other quantities may be derived and is the energy deposited in a single interaction, i, thus (see Fig. (4.1))

$$\epsilon_i = \epsilon_{\text{in}} - \epsilon_{\text{out}} + Q \quad \text{Unit : J or eV} \tag{4.1.1}$$

where ϵ_{in} is the energy of the incident ionizing particle and ϵ_{out} is the sum of the energies of all ionizing particles leaving the interaction. The rest energies are not included in ϵ. Q is the change in the rest energies of the nucleus and of all elementary particles involved in the interaction. $Q < 0$ implies an increase in rest energy, and $Q > 0$ a decrease in rest energy. Atomic excitations will give $Q=0$.

Energy imparted
In a volume there are normally many energy deposits. For this situation the quantity *energy imparted* has been defined as

$$\epsilon = \sum \epsilon_i \quad \text{Unit : J or eV} \tag{4.1.2}$$

where the summation is performed over all energy deposits, ϵ_i, in that volume.

The *mean energy imparted*, $\bar{\epsilon}$, to the matter in a volume is equal to the mean radiant energy, R_{in}, of all uncharged and charged particles entering the volume minus the mean radiant energy, R_{out}, of all uncharged and charged particles leaving the volume, plus $\sum Q$, i.e. all changes of the rest energy of nuclei and elementary particles that occur in the volume.

$$\bar{\epsilon} = R_{\text{in}} - R_{\text{out}} + \sum Q \tag{4.1.3}$$

Energy transferred

For uncharged radiation it is possible to define *energy transferred*, ϵ_{tr}, (Attix 1986) to a volume as

$$\epsilon_{tr} = (R_{in})_u - (R_{out})_u^{nonr} + \sum Q \quad \text{Unit : J or eV} \tag{4.1.4}$$

where $(R_{in})_u$, is the mean radiant energy of all uncharged particles entering the volume, $(R_{out})_u^{nonr}$ is the mean radiant energy of uncharged particles leaving the volume except which originates from radiative losses of kinetic energy before leaving the volume, plus $\sum Q$, i.e. all changes of the rest energy of nuclei and elementary particles that occur in the volume. Radiative losses are mainly bremsstrahlung and in flight annihilation by positrons.

Net energy transferred

In some situations it is practical to define *net energy transferred*, ϵ_{tr}^n, as

$$\epsilon_{tr}^n = (R_{in})_u - (R_{out})_u^{nonr} - R_u' + \sum Q = \epsilon_{tr} - R_u' \quad \text{Unit : J or eV} \tag{4.1.5}$$

where R_u' is the radiant energy emitted as radiative losses by charged particles produced in the volume independent of where radiation loss occurs.

Fig. (4.1) illustrates the concept of *energy imparted, energy transferred* and *net energy transferred* when a photon enters a volume and undergoes a pair production in the nuclear field. The initial kinetic energy of the electron is T_1, the initial kinetic energy of the positron is T_2, and T_3 is the kinetic energy of the electron when leaving the volume. The positron is annihilated in the volume and the electron undergoes a bremsstrahlung interaction outside the volume.

If the positron is annihilated at rest then

$$\epsilon = h\nu - T_3 - 2m_e c^2 + \sum Q \tag{4.1.6}$$

$\sum Q = -2m_e c^2 + 2m_e c^2 = 0$ as the production of the electron-positron pair is compensated by the annihilation of the positron.

As $h\nu = T_1 + T_2 + 2m_e c^2$ the energy imparted can be expressed as

$$\epsilon = T_1 + T_2 + 2m_e c^2 - T_3 - 2m_e c^2 = T_2 + (T_1 - T_3) \tag{4.1.7}$$

The energy transferred is

$$\epsilon_{tr} = h\nu - 2m_e c^2 + \sum Q \tag{4.1.8}$$

With $h\nu = T_1 + T_2 + 2m_e c^2$ and $\sum Q = 0$ the energy transferred ϵ_{tr} can be expressed as

$$\epsilon_{tr} = T_1 + T_2 + 2m_e c^2 - 2m_e c^2 = T_1 + T_2 \tag{4.1.9}$$

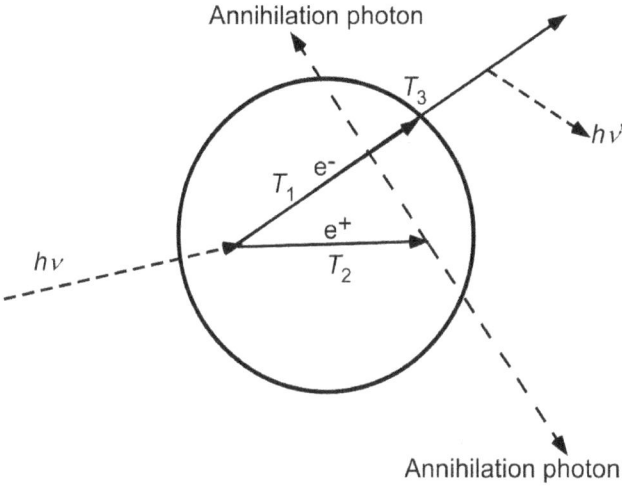

Figure 4.1: Energy transfer in a small cavity. A photon with the energy hv enters the cavity and undergoes pair production, resulting in an electron-positron pair. The positron is absorbed in the cavity and produces two annihilation photons, but the electron leaves the cavity with the energy T_3.

The energy transferred is thus the kinetic energy transferred to the charged particles in the volume.

The net energy transferred will be

$$\epsilon_{tr}^n = hv - 2m_e c^2 - hv' + \sum Q \tag{4.1.10}$$

where hv' is the energy of the bremsstrahlung photon produced outside the volume.

The net energy transferred can then be written as

$$\epsilon_{tr}^n = T_1 + T_2 + 2m_e c^2 - hv' - 2m_e c^2 = T_1 + T_2 - hv' \tag{4.1.11}$$

The net energy transferred is thus the kinetic energy transferred to the charged particles in the volume subtracted with the kinetic energy transferred to a photon.

Lineal energy

The stochastic quantity *lineal energy*, y, is the quotient of ϵ_s by $\bar{l}$, where ϵ_s is the energy imparted to matter in a volume by a single event and $\bar{l}$ is the mean chord length of that volume.

$$y = \frac{\epsilon_s}{\bar{l}} \quad \text{Unit}: J \, m^{-1} \tag{4.1.12}$$

Often the unit of lineal energy is expressed as keV μm^{-1}.

For convex volumes, the mean chord length, $\bar{l}$, is equal to 4V/S, where V is the volume and S is the surface of the volume. For a sphere this gives $\bar{l}$=4r/3 where r is the radius of the sphere.

Kerma

Kerma (kinetic energy released per mass), K, for ionizing uncharged particles is the quotient of dE_{tr} by dm, where dE_{tr} is the mean sum of the initial kinetic energies of all charged particles liberated in a mass dm of a material by the uncharged particles incident on dm

$$K = \frac{dE_{tr}}{dm} \quad \text{Unit}: J\,kg^{-1} \tag{4.1.13}$$

The special name for the unit of kerma is gray (Gy).

Kerma can be obtained through the relation

$$K = \int \Phi_E E(\mu_{tr}/\rho)\,dE = \int \Psi_E(\mu_{tr}/\rho)\,dE \tag{4.1.14}$$

where Φ_E is fluence of uncharged particles differential in energy, Ψ_E is energy fluence of uncharged particles differential in energy, E is particle energy and (μ_{tr}/ρ) is the mass energy transfer coefficient.

As kerma can be expressed in terms of a fluence, one can refer to kerma at a point in free space or inside a different material. It is thus possible to determine air kerma inside a water phantom.

Kerma can also be expressed using the energy transferred, ϵ_{tr}.

$$K = \frac{d\epsilon_{tr}}{dm} \tag{4.1.15}$$

Collision kerma

In some situations it is of interest to define a *collision kerma* which then excludes radiative losses, resulting when the charged particles are absorbed in the medium. The main part of this loss is due to bremsstrahlung, but also annihilation in flight may be of importance.

Collision kerma can be obtained through the relation

$$K_c = \int \Phi_E E(\mu_{en}/\rho)\,dE = \int \Psi_E(\mu_{en}/\rho)\,dE \tag{4.1.16}$$

where (μ_{en}/ρ) is the mass energy absorption coefficient.

Collision kerma can also be expressed using the net energy transferred, ϵ_{tr}^n

$$K_c = \frac{d\epsilon_{tr}^n}{dm} \tag{4.1.17}$$

Air-kerma-rate constant

The *air-kerma-rate constant*, Γ_δ, of a radionuclide emitting photons is the quotient of

$l^2 \dot{K}_\delta$ by A, where $\dot{K}_\delta$ is the air kerma rate due to photons of energy greater than δ, at a distance l in vacuum from a point source with activity A.

$$\Gamma_\delta = \frac{l^2 \dot{K}_\delta}{A} \quad \text{Unit}: \text{m}^2\,\text{J}\,\text{kg}^{-1} \qquad (4.1.18)$$

The special unit for the air-kerma-rate constant is $\text{m}^2\,\text{Gy}\,\text{Bq}^{-1}\,\text{s}^{-1}$.

Cema

For ionizing charged particles a quantity corresponding to kerma for uncharged particles is *cema* (converted energy per mass). The cema, C, is the quotient of $\mathrm{d}E_{\text{el}}$ by $\mathrm{d}m$, where $\mathrm{d}E_{\text{el}}$ is the mean energy lost in electronic interactions in a mass $\mathrm{d}m$ of a material by charged particles, except secondary electrons, incident on $\mathrm{d}m$.

$$C = \frac{\mathrm{d}E_{\text{el}}}{\mathrm{d}m} \quad \text{Unit}: \text{J}\,\text{kg}^{-1} \qquad (4.1.19)$$

The energy lost by secondary electrons is not included in $\mathrm{d}E_{\text{el}}$, but contribution from all other charged particles as secondary protons and α-particles are included.

Cema can be obtained through the relation

$$C = \int \Phi_E (S_{\text{el}}(E)/\rho)\,\mathrm{d}E = \int \Phi E (L_\infty(E)/\rho) \qquad (4.1.20)$$

where Φ_E is fluence of charged particles of energy E, S_{el}/ρ is mass collision (electronic) stopping power and L_∞ is the unrestricted linear energy transfer.

In some situations it is of interest to disregard the transport of the secondary electrons and include the energy lost by them. In this case *restricted cema* may be used, defined as

$$C_\Delta = \Phi'_E L_\Delta / \rho \qquad (4.1.21)$$

where Φ'_E includes secondary electrons with kinetic energies larger than Δ. When $\Delta = \infty$ then restricted cema is equal to cema.

Absorbed dose

The *absorbed dose*, D, is the quotient of $\mathrm{d}\bar{\varepsilon}$ by $\mathrm{d}m$, where $\bar{\varepsilon}$ is the mean energy imparted by ionizing radiation to matter in a volume of mass $\mathrm{d}m$

$$D = \frac{\mathrm{d}\bar{\varepsilon}}{\mathrm{d}m} \quad \text{Unit}: \text{J}\,\text{kg}^{-1} \qquad (4.1.22)$$

The special unit for absorbed dose is gray (Gy). The absorbed dose, D, is considered as a point quantity, but the physical process does not allow the mass to go to zero. However, sometimes the absorbed dose is written as

$$D = \lim_{m \to 0} \bar{z} \qquad (4.1.23)$$

This is however not formally correct.

4.1.2 Radiation Equilibra

To accurately calculate the absorbed dose in a material, a complete picture of the fluence differential in both energy and angle is necessary at all points in it, together with information of all the subsequent interaction processes. A statistical representation of this can be obtained by using Monte Carlo calculation methods. However, in some situations simplified approximations may be used. In this context the following equilibrium situations will shortly be discussed: Radiation equilibrium, charged particle equilibrium, delta-particle equilibrium and partial delta-particle equilibrium.

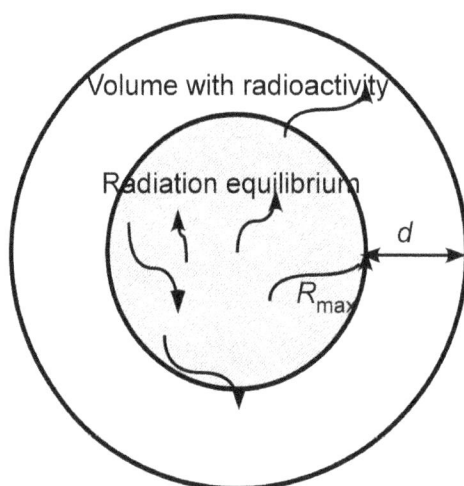

Figure 4.2: The full volume is filled with a uniform radioactive solution. Radiation equilibrium exists in the partial volume at distances which are larger than the maximum range, R_{max}, from the surface, d, of the emitted particles.

Radiation equilibrium

If the radiant energy R_{in} is equal to R_{out} then the energy imparted can be calculated according to Eq. (4.1.3)

$$\bar{\epsilon} = R_{in} - R_{out} + \sum Q = \sum Q \qquad (4.1.24)$$

This gives that the absorbed dose is

$$D = \frac{d\bar{\epsilon}}{dm} = \frac{d\sum Q}{dm} \qquad (4.1.25)$$

Consider a large volume filled uniformly with radioactivity (See Fig. 4.2). If the maximum distance from a point to the border of the volume is larger than the range of the emitted particles in the decay and their secondaries, then radiation equilibrium exists at that point. The absorbed dose at this point is given by

$$D = \frac{dS}{dm}\bar{T} \qquad (4.1.26)$$

where dS/dm is the number of ionizing particles emitted per mass unit in the medium and $\bar{T}$ is the mean kinetic energy of the emitted particles including the kinetic energy of the daughter nuclide. The neutrinos shall not be included in the calculations as their interaction can be neglected.

The conditions for radiation equilibrium are most easily obtained for a pure alpha-emitter, without any following photons, as the range of the alpha-particles in tissue is less than 0.1 mm. Beta-radiation has practical ranges less than about one cm, and for larger volumes, it is possible to obtain radiation equilibrium centrally, if production of bremsstrahlung is neglected. However, for gamma-radiation it is not possible to get radiation equilibrium as the photons are attenuated exponentially.

At the surface of the radioactive volume, the absorbed dose is approximately half of that in the part where there is radiation equilibrium.

Charged particle equilibrium

If in a volume the radiant energy from charged particles entering the volume equals

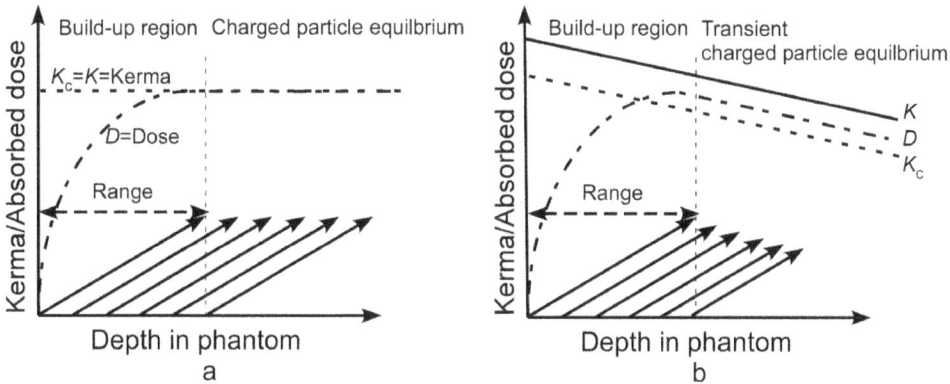

Figure 4.3: Photons are impinging from left on a phantom. Electrons emitted are illustrated by the arrows. In **a)** the number of electrons emitted per length unit is constant and there is a situation with charged particle equilibrium after a depth corresponding to the range of the electrons. In **b)** the attenuation of photons has been considered and the number of electrons produced decreases with depth. In this situation there is transient charged particle equilibrium.

the radiant energy leaving the volume, then there is charged particle equilibrium (CPE). Consider the situation illustrated in Fig. 4.3, where a clean photon beam impinges on matter from the left. The photons will emit electrons through interactions in the medium. If the approximation is made that the attenuation of the photon beam over the range of the electrons can be neglected, there will be charged particle equilibrium at depths larger than the range of the electrons. This approximation holds best for low energy photons and can be used with better accuracy than one per cent for energies below 500 keV. For higher energies this approximation fails and this situation is referred to as transient charged particle equilibrium (TCPE). In Fig. 4.3, the variation of the absorbed dose, collision kerma, and kerma are illustrated. For the situation with CPE, collision kerma and absorbed dose are the same. Thus when CPE holds the absorbed dose can be obtained from

$$D = K_c = \int \Phi_E E (\mu_{en}/\rho) \, dE = \int \Psi_E (\mu_{en}/\rho) \, dE \qquad (4.1.27)$$

At low photon energies where CPE holds, then also the difference between kerma and collision kerma is small and thus

$$D = K_c = K \qquad (4.1.28)$$

In the situation with TCPE the absorbed dose is larger than the collision kerma. Kerma is also larger than the collision kerma at high energies as the probability for radiative losses of the emitted electrons can not be neglected (Fig. 4.3 b).

Delta-particle equilibrium

In a charged particle beam there is no CPE, but it is possible to obtain delta-particle equilibrium (DPE) if the range of the delta-particles is small compared to the range of the primary charged particles. This can be a good approximation in a heavy charged particle beam, as in a proton beam, because the energy transferred to the delta-particles is very small. When DPE exists the absorbed dose is given as

$$D = \int \Phi_{E,P} (S_{el}(E)/\rho) \, dE = \int \Phi_{E,P} (L_\infty(E)/\rho) \, dE \qquad (4.1.29)$$

where $\Phi_{E,P}$ is the fluence of primary charged particles differential in energy E and $S_{el}(E)/\rho$ is the mass collision stopping power for particles with the energy E.

Partial delta-particle equilibrium

In an electron beam it is difficult to obtain DPE as the electrons can transfer a large part of their energy to the delta-particles. However, it is then possible to define a partial delta-particle equilibrium for energies below a certain cut-off energy, typically around 100 eV. The absorbed dose can then for mono-energetic electrons be simplified to

$$D = \int \Phi_{\Delta,E} (S_{el,\Delta}(E)/\rho) \, dE = \int \Phi_{\Delta,E} (L_\Delta(E)/\rho) \, dE \qquad (4.1.30)$$

where $\Phi_{\Delta,E}$ is the differential fluence of the primary particles and delta-particles with energy above Δ. $S_{el,\Delta}(E)/\rho$ is the restricted mass collision stopping power for energy transfers below Δ.

These relations can be compared to the quantities *cema* and *restricted cema*, see above.

4.1.3 Cavity Theories

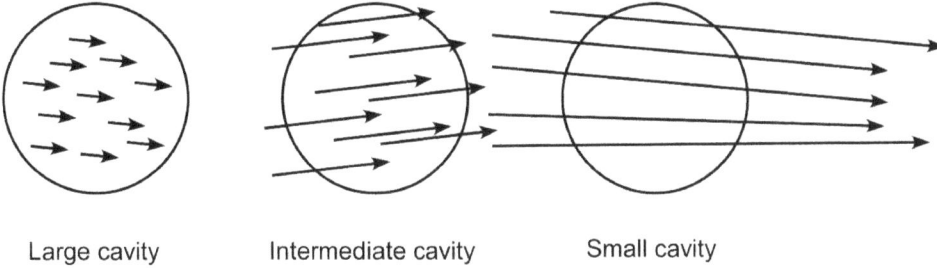

Large cavity   Intermediate cavity   Small cavity

Figure 4.4: In the left cavity the range of the electrons is small, in the central cavity the electron range is of the same order, and in the right cavity the electron range is large as compared to the size of the cavity.

In order to determine the absorbed dose experimentally a detector, or a "cavity", is introduced into the medium. This cavity can also be just a part of the medium with a different composition and density, but here it will mainly be discussed from the detector point of view. In a photon beam there are mainly three different situations depending on the size of the cavity in comparison to the range of the emitted electrons; large cavity, small cavity, and intermediate cavity (See Fig. 4.4).

Large cavity
A photon beam hits a medium and assuming that CPE exists in the medium, then the absorbed dose in the medium is given by (compare Eq. (4.1.27))

$$D_{\text{med}} = \int \Psi_{E,\text{med}}(\mu_{\text{en}}/\rho)_{\text{med}} \, \mathrm{d}E \qquad (4.1.31)$$

A detector (cavity) is now positioned at the point where the absorbed dose shall be determined. Assuming that there is CPE also in the detector, as it is large compared to the range of the electrons (See Fig. 4.4), the mean absorbed dose to the detector is

$$\bar{D}_{\text{det}} = \int \bar{\Psi}_{E,\text{det}}(\mu_{\text{en}}/\rho)_{\text{det}} \, \mathrm{d}E \qquad (4.1.32)$$

The ratio of the absorbed doses is given by

$$\frac{D_{\text{med}}}{\bar{D}_{\text{det}}} = \frac{\int \Psi_{E,\text{med}}(\mu_{\text{en}}/\rho)_{\text{med}} \, \mathrm{d}E}{\int \bar{\Psi}_{E,\text{det}}(\mu_{\text{en}}/\rho)_{\text{det}} \, \mathrm{d}E} \qquad (4.1.33)$$

Assuming that the detector does not disturb the photon energy fluence and thus $\Psi_{E,\text{med}}$ and $\bar{\Psi}_{E,\text{det}}$ are the same, this can be written as

$$\frac{D_{\text{med}}}{\bar{D}_{\text{det}}} = (\bar{\mu}_{\text{en}}/\rho)_{\text{med,det}} \qquad (4.1.34)$$

This relation then neglects the small transfer region in the detector close to the medium, where the absorbed dose is changing from that in the medium to in the detector.

Small cavity

With a chamber that is small compared to the range of the electrons it is possible to use the Bragg-Gray cavity theory. This theory was derived for a gas-filled cavity in a medium irradiated with photons, but it can also be extended to other types of cavities.

The following conditions should hold:
1. Introducing a small gas-filled cavity into the medium will not change the fluence of the charged particles. This assumption is based on Fano's theorem.
2. Photon interactions in the cavity can be neglected.
3. The photon fluence is uniform over the cavity volume.

If these conditions hold, then the primary electron fluence can be assumed to be the same in the medium and in the detector and the ratio between the absorbed dose in the medium and the detector (cavity) is given by

$$\frac{D_{\mathrm{med}}}{\bar{D}_{\mathrm{det}}} = \frac{\int \Phi_E (S(E)_{\mathrm{el}}/\rho)_{\mathrm{med}} \, dE}{\int \bar{\Phi}_E (S(E)_{\mathrm{el}}/\rho)_{\mathrm{det}} \, dE} \tag{4.1.35}$$

This can be written as

$$\frac{D_{\mathrm{med}}}{\bar{D}_{\mathrm{det}}} = (\bar{s}_{\mathrm{el}}/\rho)_{\mathrm{med,det}} \tag{4.1.36}$$

This equation holds only if there is delta-particle equilibrium, i.e. the energy lost by interactions in the cavity is also locally absorbed in the cavity. Spencer and Attix extended the Bragg-Gray cavity to take into consideration the delta-particles. All electrons, including delta-particles with energies above a cut-off energy Δ, were included in the fluence spectrum, while electrons with energies below Δ were assumed to be locally absorbed in the cavity. The electron energy with a range compared to the size of the cavity was used as a basis for the choice of Δ. For a typical ionization chamber, the value is often set to Δ=10 keV. The ratio of doses according to Spencer-Attix is expressed as

$$\frac{D_{\mathrm{med}}}{\bar{D}_{\mathrm{det}}} = \frac{\int_{\Delta}^{E_{\mathrm{max}}} \Phi_E^{\delta} (L_{\Delta}(E)/\rho)_{\mathrm{med}} \, dE + \left[\Phi_E(\Delta)(S_{\mathrm{el}}(\Delta)/\rho)_{\mathrm{med}}\Delta\right]}{\int_{\Delta}^{E_{\mathrm{max}}} \bar{\Phi}_E^{\delta} (L_{\Delta}(E)/\rho)_{\mathrm{det}} \, dE + \left[\bar{\Phi}_E(\Delta)(S_{\mathrm{el}}(\Delta)/\rho)_{\mathrm{det}}\Delta\right]} \tag{4.1.37}$$

The factors in the bracket take into consideration the so-called track-ends (Nahum, 1978), which correct for the energy deposition of the electrons falling below the energy Δ.

To calculate these stopping power ratios, Monte Carlo methods are typically used. However, in many situations where the difference in atomic number of the medium and the cavity is small, mean values can be used as a good approximation. In the compilations in this material it is assumed that the mean value of the stopping power ratio can be obtained using the following approximations:

1. Assume a mean value for the photon spectrum, if necessary.
2. Calculate a mean value of the emitted electron energy by using the interaction coefficient for each process; photoelectric effect, incoherent scattering and pair production in the nuclear and electron field. Determine the mean electron energy for each process and calculate the total mean value, taking into consideration the number of electrons emitted in each interaction.
3. The obtained mean value is the energy of the emitted electrons. When passing through the medium the energy is decreased. Different assumptions of the final mean value have been used. Attix (Attix, 1986) has proposed to multiply with a factor of 0.5 and ICRU with a factor of 0.4 (ICRU, 1969). In the proposed solutions in this material the factor of 0.5 has been used.

Intermediate cavity

A general cavity theory should cover all type of cavities from small to large. Several approaches have been used throughout the years and the most popular has been the sometimes called Burlin theory (Burlin, 1966). In a simplified expression it can be written as

$$\frac{\bar{D}_{det}}{\bar{D}_{med}} = d(\bar{s}_{el}/\rho)_{det,med} + (1-d)(\bar{\mu}_{en}/\rho)_{det,med} \qquad (4.1.38)$$

This relation is based on the assumption that the cavity will modify the electron fluence in two ways:

1. The relative electron energy spectrum in the cavity is changed as the stopping power of the cavity is different from the stopping power in the medium.
2. The total energy in the electron spectrum is changed as the absorption of photons is different in the cavity and the medium.

The weighting factor d should be equal to unity for small cavities and equal to zero for large cavities. The expression for calculating, d, proposed by Burlin is

$$d = \frac{1-e^{-\beta L}}{\beta L} \qquad (4.1.39)$$

where β is the absorption coefficient of the electrons in the cavity but emitted in the medium as well in the cavity. When L, the mean chord length in the cavity, approaches zero, d tends to unity; and when L approaches ∞, then d tends to zero. This expression for the Burlin theory will agree with the expressions above for large and small cavities. The value of d proposed by Burlin is rather approximate and several approaches to improve it have been published. However, the difference between the various newer approaches is small. The need for this theory has become of less interest with the use of Monte Carlo and only one example is included in the exercises in this compilation.

4.1.4 Ionization Chamber Dosimetry

In radiotherapy, ionization chamber measurements are the main way used to determine the amount of an absorbed dose. The procedure is described in different protocols like IAEA TRS 398 (IAEA, 2000) and AAPM TG 51 (Almond et al, 1999). The description below follows the IAEA TRS 398 protocol. Only some relations of importance for solving the exercises in this compilation are included. For more information see the full protocol.

In the discussion it is assumed that the ionization chamber is calibrated at a Secondary or Primary Standard Laboratory where the reference quality, Q_0, is a ^{60}Co gamma ray beam and a calibration factor in terms of absorbed dose to water, N_{D,w,Q_0}, is obtained, often referred to as $N_{D,w}$ when Q_0 refers to a ^{60}Co gamma ray beam.

Determination of the absorbed dose to water in the user beam
The absorbed dose to water under reference conditions is given by

$$D_{w,Q} = M_Q N_{D,w} k_{Q,Q_0} \tag{4.1.40}$$

where M_Q is the reading of the ionization chamber corrected for influence quantities and k_{Q,Q_0} is the chamber specific correction factor or k_Q for a ^{60}Co calibration beam, which corrects for the difference in radiation quality between the reference quality Q_0 and the user quality Q.

When the user quality is the same as the reference quality, k_{Q,Q_0} is equal to unity. The absorbed dose to water under reference conditions is then given by

$$D_w = M N_{D,w} \tag{4.1.41}$$

Correction factors
The beam quality factor k_{Q,Q_0}, is defined as the ratio of the calibration factor at the radiation qualities, Q_0 and Q

$$k_{Q,Q_0} = \frac{N_{D,w,Q}}{N_{D,w,Q_0}} = \frac{D_{w,Q}/M_Q}{D_{w,Q_0}/M_{Q_0}} \tag{4.1.42}$$

When no calibration is possible for different radiation qualities, the correction factor can be obtained from the relation (IAEA, 2000)

$$k_{Q,Q_0} = \frac{(\bar{s}_{w,air})_Q \left(\bar{W}_{air}\right)_Q p_Q}{(\bar{s}_{w,air})_{Q_0} \left(\bar{W}_{air}\right)_{Q_0} p_{Q_0}} \tag{4.1.43}$$

The values of the Spencer-Attix water-air stopping power ratios, $\bar{s}_{w,\text{air}}$, as well as the mean energy expended in air per ion pair produced, $\bar{W}_{\text{air}}$, should be averaged over the whole particle spectrum. This holds also for the perturbation factors p_Q, which correct for any perturbation of the particle fluence produced by inserting the chamber into the phantom. The total perturbation factor p_Q includes p_{wall}, p_{cav}, p_{dis} and p_{cel}. The perturbation factor is the only chamber dependent factor in the equation.

For photon and electron beams the variation of $\bar{W}_{\text{air}}$ with energy is small and can be neglected. As such the equation for k_{Q,Q_0} can be reduced to

$$k_{Q,Q_0} = \frac{(\bar{s}_{w,\text{air}})_Q \, p_Q}{(\bar{s}_{w,\text{air}})_{Q_0} \, p_{Q_0}} \qquad (4.1.44)$$

Correction factors k_{Q,Q_0} are tabulated in IAEA TRS 398 (IAEA, 2000); Table 14 for high energy photon beams and Table 18 for electron beams.

Influence quantities
The measured value from the ionization chamber must be corrected for conditions different from the reference conditions. The general influence quantities are discussed shortly below.

Correction for temperature and pressure
An ionization chamber to be used for calibration is recommended to be open to the ambient atmosphere. As the absorbed dose is related to the mass in the chamber and the density of air is dependent on temperature and pressure; a correction has to be applied according to

$$k_{Tp} = \frac{Tp_0}{T_0 p} \qquad (4.1.45)$$

where T is the temperature in K and p is the air pressure in kPa at measurement at the user and T_0 is the temperature in K and p_0 the air pressure at the calibration. Often T_0 is 293.2 K and p_0 is 101.3 kPa.

Correction for humidity
If the calibration was made at a relative humidity of 50% and the humidity at the measurement is between 20% and 80%, no correction is necessary. If the calibration factor is referring to a humidity of 0% then a small correction factor should be applied.

Correction for polarity effect
When changing the polarity of the chamber, different readings can be obtained. This must be checked and if necessary, corrected for. If there is a measurable polarity effect, the correction factor is obtained from

$$k_{\text{pol}} = \frac{|M_+| + |M_-|}{2M} \qquad (4.1.46)$$

where M_+ and M_- are the electrometer readings for positive and negative polarity and M is the electrometer reading for the polarity normally used in the measurements. The use of this factor assumes that the polarity correction has been applied at calibration.

Correction for ion recombination
All ions produced in the chamber may not be collected at the electrodes. There are mainly two separate effects that take place:
1. Recombination of ions within a single particle track, referred to as initial recombination. This effect is independent of dose rate but is radiation quality dependent.
2. Recombination of ions from separate particle tracks, referred to as general or volume recombination. This effect is dependent on dose rate. In pulsed beams and particular in pulsed scanned beams the ion density in the pulse may be high, and the recombination significant. Both effects are dependent on chamber geometry and applied voltage.

In practical dosimetry the recombination is often determined using the two-voltage method. Measuring at two different applied voltages give two different values of measured ionization. Assuming a linear dependence between 1/M and 1/V the correction factor for pulsed beams from an electron accelerator can be obtained from IAEA TRS 398(IAEA, 2000).

$$k_s = a_0 + a_1 \left(\frac{M_1}{M_2} \right) + a_2 \left(\frac{M_1}{M_2} \right)^2 \tag{4.1.47}$$

M_1 and M_2 are the collected charges at the polarizing voltages V_1 and V_2 respectively. The values of the constants a_i are given in Table 9 IAEA TRS 398 (IAEA, 2000) for pulsed and pulsed scanned beams.

In continuous radiation, the relation for the two voltage method is given by

$$k_s = \frac{(V_1/V_2)^2 - 1}{(V_1/V_2)^2 - (M_1/M_2)} \tag{4.1.48}$$

4.1.5 Calorimetric and Chemical Dosimeters

Besides ionization chambers two types of dosimeters have been considered accurate for calibration in radiotherapy, the *calorimeter* and the *ferrous sulfate dosimeter*.

Calorimeter
When ionizing radiation is absorbed in matter, the energy is obtained as heat. Thus by measuring the increase in temperature, it is possible to determine the absorbed energy or absorbed dose. However, the absorbed doses normally used in radiotherapy will result in a very small temperature increase. An absorbed dose of 1 Gy, will result in a temperature increase in water of about 0.24 mK. Thus calorimetry is mainly used at standard laboratories where a very well defined temperature control is possible.

The dose to water is given by

$$D_{\text{water}} = c\Delta T(1 + D_T) \tag{4.1.49}$$

where c is the specific heat capacity for water ($4.1815 \cdot 10^3$ J kg^{-1} K^{-1}), ΔT is the increase in temperature and D_T is the thermal effect. The thermal effect is defined as

$$D_T = \frac{E_a - E_h}{E_a} \tag{4.1.50}$$

where E_a is the energy imparted to water and and E_h is the energy appearing as heat.

The thermal effect also called heat defect is negligible in pure elements like carbon, but may be up to some percent for different plastic materials and water.

Ferrous sulfate dosimeter

The chemical reactions produced by ionizing radiation can be used for determining absorbed dose. The most common dosimeters are water based, and is dominated by the ferrous sulfate dosimeter, also called the Fricke dosimeter. This dosimeter uses the oxidation of Fe^{2+} ions to Fe^{3+}. The absorbed dose to the dosimeter solution is given by

$$\bar{D} = \frac{\Delta M}{\rho G(Fe^{3+})} \tag{4.1.51}$$

where ΔM is change in the molar concentration of Fe^{3+} ions (mol dm^{-3}), ρ is density of dosimeter solution (1.024 kg dm^{-3}), $G(Fe^{3+})$ is the radiation chemical yield, i e the number of moles (Fe^{3+}) produced per absorbed energy (mol/J).

The measurement of ΔM is made using a spectrophotometer, where the absorbance at the wavelength 304 nm is typically used. The absorbed dose is then given by

$$\bar{D} = \frac{\Delta(OD)}{\rho \epsilon l G(Fe^{3+})} \tag{4.1.52}$$

where OD is obtained from the relation

$$\frac{I}{I_0} = 10^{-\Delta(OD)} \tag{4.1.53}$$

and

$$\Delta(OD) = \epsilon l \Delta M \tag{4.1.54}$$

where I/I_0 is the ratio of light intensity after and before transmission through the dosimeter sample. ϵ is the molar extinction coefficient (2187 dm^3 mol^{-1} cm^{-1} at 304 nm and 25°C), l is the light path length (10 mm for a typical spectrophotometer). $G(Fe^{3+})$ is $1.606 \cdot 10^{-6}$ mol J^{-1} for ^{60}Co gamma rays.

As ϵ and $G(Fe^{3+})$ are related it is often common to determine the product. For electron energies between 1 and 30 MeV a value of $352 \cdot 10^{-6}$ m^2 kg^{-1} Gy^{-1} is recommended (ICRU, 1984), and can also be used for ^{60}Co gamma rays. The value of ϵ is temperature dependent and normally the value is given for a temperature of 25°C. For other temperatures a correction by increasing ϵ with 0.7% per °C should be applied.

4.2 Exercises in Dosimetry

4.2.1 Definitions and Important Quantities

Exercise 4.1. What energies shall be included in "energy imparted", ϵ, "net energy transferred", ϵ_{tr}^n, and "energy transferred", ϵ_{tr} in Fig. 4.5 if the positron is annihilated in flight with a kinetic energy of 0.10 MeV?

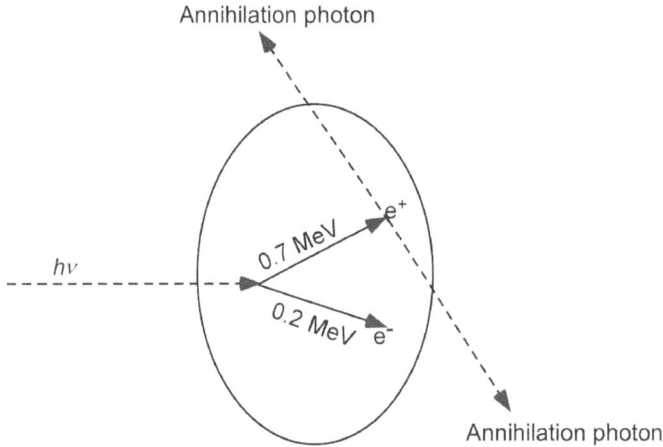

Figure 4.5: Illustration of a pair production process producing an electron-positron pair, followed by absorption of the positron through annihilation inside the cavity.

Exercise 4.2. 20 MeV electrons are impinging perpendicularly on a water phantom. Measurements are made at different depths in water with three different detectors: A) Ionization chamber, B) Faraday detector, (which measures the charge hitting the detector) and C) Scintillation detector (total absorbing). The obtained result is shown in Fig. 4.6. Explain the shape of the different curves.

4.2.2 Radiation Equilibra

Exercise 4.3. At a point in a water phantom, there is the same fluence of photons and electrons, both with the same energy of 1.0 MeV. If charged particle equilibrium holds for the photons and δ-particle equilibrium holds for the electrons, calculate the ratio of the absorbed doses from electrons and photons respectively.

Exercise 4.4. A spherical volume in the brain, filled with liquid, has tumor

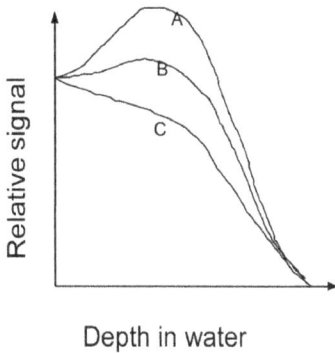

Figure 4.6: Variation of relative signal with depth for 20 MeV electrons in water obtained with A: Ionization chamber B: Faraday cup and C: Scintillation detector.

cells on the inside of the wall. These tumor cells shall be treated by filling the volume with a radioactive solution containing ^{32}P. Calculate the necessary activity to give an absorbed dose to the wall during the first two days equal to 10.0 Gy? The radius of the volume is 15 mm. What is the absorbed dose at the center of the volume? The solution may be considered as water equivalent from an absorption point of view.

Exercise 4.5. In radiotherapy with protons, the range is often modulated by positioning disks of PMMA in the beam. It is then important that there is a possibility to check if the disk is positioned in the beam or not. A diamond detector is positioned on the skin of the patient. Calculate the ratio of the signal from the diamond detector with and without a range modulating PMMA disk. Make the calculation for the following situation. The energy of protons directly after the accelerator is 85 MeV. The distance between the accelerator and the patient is 100 cm. The PMMA disk is placed 50 cm from the patient and its thickness is 15 mm. The diamond detector may be considered small in relation to the range of the protons but large in relation to the range of the δ-particles. The detector is assumed to be made of carbon only. The temperature and pressure of the air is 295 K and 101 kPa, respectively.

Exercise 4.6. Radiotherapy with "light ions" has become of interest during the last years. One advantage with this modality of treatment is that it is possible to, in a more well defined way, treat only the tumor and reduce the dose to the healthy tissue. But then it is also more critical to hit the target correctly. When calculating the absorbed dose it is also necessary to include the energy of the particles that changes with depth. Consider the following treatment situation. The particles in the beam are 150 MeV protons at the surface of the patient. The target center is situated at a depth of 12.0 cm in soft tissue, with a density of $1.0 \cdot 10^3$ kg m^{-3}. However, due to movements of the intestine and mistreatment, the beam will instead pass through 6.0 cm of soft tissue, 2.0 cm air (in the intestine), 1.0 cm bone tissue (density $1.8 \cdot 10^3$ kg m^{-3}) and 3.0 cm soft tissue. The planned absorbed dose to the target center is 2.0 Gy. Calculate the absorbed dose in the situation with non uniformity.

Exercise 4.7. A 0.40 GBq ^{139}Ce point source is positioned in a large water phantom. Calculate the absorbed dose to water from γ-radiation 100 mm from the radiation source. Assume that the relative fluence contribution from the secondary photons is given by $(\sigma_s/\sigma_a)\mu r$, where r is the distance from the radiation source to the measuring point. σ_s and σ_a are the cross sections for Compton scattering and absorption respectively. The secondary photons are assumed to be once scattered Compton photons. The mean energy may be used in the calculations.

Exercise 4.8. A radioactive source was lost from a laboratory. Somebody found the radioactive source and took it home without understanding that the object was radioactive. The source was put on a table in the home. Later a film, that had been stored close to the source, was developed. It was found to have a strong background density that destroyed the real picture. It was understood that the reason probably was irradiation by ionizing radiation and the object was now discovered to be radioactive. Calculate the absorbed dose to the film using the following assumptions. The radioactive source was a ^{137}Cs source with the activity 42.5 MBq. The irradiation time was 300 h with a distance between the film and the source of 20.0 cm. The absorbed dose is obtained by calculating the absorbed dose to a small mass element in the film emulsion with a size that approximately gives charged particle equilibrium. The radioactive source may be assumed to be a point source and the attenuation in air and film cover neglected.

Exercise 4.9. When treating goiter or cancer in the thyroid, the patient is given a solution containing radioactive ^{131}I. This is mainly taken up in the thyroid, that thus becomes a radioactive source. An important question is then if the patient can stay together with other people or will they then be irradiated and obtain large absorbed doses. Make an estimate of the absorbed dose using these simplified assumptions: The activity in the thyroid is 2.50 GBq. The thyroid may be considered as a point source, positioned 10 mm below the skin. Assume no excretion of the activity, which means that the dose rate is decreasing with the radioactive half life. Calculate the absorbed dose to a small mass of tissue free in air, situated at a point 50.0 cm from the skin, during the first 10.0 h after the intake, assuming the uptake is instantaneous. Only primary photons with the most frequent energy need be included in the calculations. Assume a tissue density of $1.0 \cdot 10^3$ kg m^{-3}.

4.2.3 Cavity Theories

Exercise 4.10 In connection to a nuclear power accident, milk can be contaminated with ^{90}Sr/^{90}Y. To determine the activity in the milk, a LiF-dosimeter is placed at the center of a container with the volume 100 dm^3 and filled with milk. The LiF-dosimeter is left at that position for a week. The measurement shows that the dosimeter has obtained an absorbed dose of 1.85 mGy. Calculate the activity of ^{90}Sr/^{90}Y. From a dose calculation point of view, the milk can be regarded as equivalent to water. Your result should show that the milk should not be used for drinking from radioactivity point of view.

Exercise 4.11 In a measurement, LiF dosimeters were placed in a water phantom and irradiated with γ-radiation from ^{99m}Tc. The LiF-dosimeters have a diameter of 13 mm and a thickness of 0.10 mm. Their density is 2.50·10^3 kg m^{-3}. The projected range of the electrons may be considered as small compared to the thickness of the dosimeter. To protect the dosimeter from water, they are placed in a container made of polystyrene with a wall thickness of 1.0 mm, which may be assumed to be large compared to the projected range of the electrons, but small in comparison to the mean free path of the photons. The absorbed dose to water at the point where the dosimeters are positioned shall be 0.10 Gy. How many measurement units (m.u.) are needed if the calibration of the LiF-dosimeters with ^{60}Co γ-radiation resulted in 37200 m.u. per Gy in water. When irradiated with ^{60}Co γ-radiation Bragg-Gray relation may be assumed to hold. Neglect contribution from secondary photons.

Exercise 4.12 When determining absorbed dose to water it is often important to have detectors that are as water equivalent as possible. At the same time it is important to have detectors with small dimensions. In a measurement of narrow photon beams there is a possibility to choose between a silicon diode and a diamond detector. The detectors are irradiated at a point in a water phantom, where previously the absorbed dose to water have been determined to be 2.00 Gy. Calculate the absorbed dose to the detectors if they can be assumed to be
a) large compared to the electron range
b) small compared to the electron range
The radiation source is a linear accelerator with a photon mean energy of 4.0 MeV.

Exercise 4.13 In connection with radiotherapy one often measures the absorbed dose to water. For practical reasons, sometimes a plastic material is used instead. Calculate the ratio of absorbed doses to a small LiF-dosimeter, that is placed in a water phantom and a polystyrene phantom, respectively. The dosimeter is placed at dose maximum in the two materials and one can assume that the same photon fluence and energy distribution is at hand at the two different dose maxima. A 6 MV linear accelerator is used as radiation source, with a mean energy of 2.5 MeV. The

LiF dosimeter is assumed to be a Bragg-Gray cavity. Assume that charged particle equilibrium holds at the dose maximum in the two phantoms.

Exercise 4.14 A LiF-dosimeter with a thickness of 0.50 mm and a diameter of 13 mm, is positioned in water and irradiated with 10 MV x rays with a mean energy of 4.0 MeV. The absorbed dose to the dosimeter is determined to 1.47 Gy. Calculate the absorbed dose to water assuming that the Burlin theory can be used. The density of the LiF dosimeter is $2.30 \cdot 10^3 \, \text{kg m}^{-3}$.

Exercise 4.15 In radiotherapy it is important that the absorbed dose to the patient is correct. The dosage is based on the monitor chamber positioned in the treatment head of the accelerator. This monitor has to be calibrated to the absorbed dose in a water phantom. The normalization depth is typically 100 mm. However, the absorbed dose in the water phantom varies for the same monitor number with relation to the field size, as there are more secondary photons with a larger field. Because of this, output factors are determined, i.e. the absorbed dose as a function of the field size normalized to a 10x10 cm² field. When measuring these output factors with diodes, different factors are obtained depending on the size of the diode. The reason for this can be due to a cavity problem, i.e. the relation between the absorbed dose to water and to the diode is dependent on if the diode can be regarded as small or large compared with the range of the electrons as the secondary photons have a lower energy and this relation may vary with field size. Make an estimation of this effect using the following assumptions: The primary beam is 18 MV x rays with a mean energy of 6.0 MeV. The secondary photons are assumed to have a mean energy of 200 keV. Calculate the ratio between the absorbed dose to the dosimeter and the absorbed dose to water for the primary photons assuming that the detector is small compared to the range of the electrons and for the secondary photons assuming the detector is large as compared to the range of the electrons. The detector is regarded to be pure silicon.

Exercise 4.16 A diamond detector has a water-protecting cover of polystyrene. Calculate the ratio of the detector signal with the protecting cover compared to a diamond detector without the protecting cover, when the detector is placed at a depth of 50 mm in a water phantom. The detector may be considered as a Bragg-Gray cavity and one can assume that with the polystyrene cover, all electrons reaching the detector are produced in the cover. The radiation source is an 18 MV x-ray beam from a linear accelerator with a mean energy of 5.0 MeV at the measuring point.

Exercise 4.17 A polystyrene phantom is irradiated with 40 MV photons which have a mean energy of 13 MeV. At a depth of 50 mm, LiF-dosimeters are used with a diameter 13 mm, and the thickness 0.10 mm. A signal of 42300 m.u. is obtained during an irradiation of 180 s. Calculate the absorbed dose rate in polystyrene at the measuring point.

The LiF-dosimeters have been calibrated with ^{60}Co-gamma radiation in a small Teflon phantom. The collision kerma rate in air at the phantom is 0.63 Gy min^{-1}. An irradiation duration of 120 s results in a signal of 37500 m.u. The Teflon may be considered equivalent with LiF from an absorption point of view. The correction factor for the attenuation in the Teflon wall is 0.99.

Exercise 4.18. The depth dose distribution in water for an electron beam with the mean energy at the surface of 20 MeV shall be determined. The measurements are made with both a plane parallel ionization chamber and a silicon diode. Calculate the ratios of the measured values, at the depth of 80 kg m^{-2} where the dose to water is 50% of the absorbed dose at its maximum and at the depth of dose maximum of 30 kg m^{-2}.

4.2.4 Ionization Chamber Dosimetry

Exercise 4.19. The monitor for a linear accelerator is to be calibrated by using a NE 2571 Farmer ionization chamber. The ionization chamber is placed at a depth of 50 mm in a water phantom and irradiated with 4 MV x rays. Geometry: Field size=100x100 mm^2. SSD=800 mm. After an irradiation of 120 s the monitor chamber shows a value of 3.00 Gy and a value of 213.3 nC is obtained by the calibration chamber. Pressure: 100 kPa, temperature: 295 K. TPR$_{20,10}$=0.62.

To determine the recombination, the NE 2571 Farmer chamber is irradiated using the applied voltages of 300 V and 100 V. The ratio of the measured values at the two voltages is 1.012.

Calculate the monitor calibration factor if the monitor is intended to give the absorbed dose at a depth of 50 mm and a field size of 100x100 mm^2.

The calibration factor of the ionization chamber $N_{D,w}$=0.0139 Gy nC^{-1} (101.3 kPa, 293 K).

Exercise 4.20. A water phantom is irradiated with an electron beam with R_{50}=8.0 g cm^{-2}. A Capintec PR06C ionization chamber is placed at the reference depth. The calibration constant obtained in a ^{60}Co calibration of the chamber $N_{D,w}$ = 9.17 · 10^{-3} Gy nC^{-1}, at 293 K and 101.3 kPa. After an irradiation for 120 s, 235 nC is obtained. The temperature and pressure values measured are 295 K and 98.3 kPa, respectively. To determine the recombination the Capintec chamber is irradiated with applied voltages of 400 V and 100 V. The ratio of the measured values is 1.010. Calculate the absorbed dose rate to water.

Exercise 4.21. In radiation protection measurements, ionization chambers filled with gas under high pressure are often used to increase the sensitivity. An ionization chamber has walls made of aluminum and is filled with argon to a pressure of 800 kPa at temperature 293 K. The chamber volume is 50.0 cm^3. The chamber wall is thick enough to give charged particle equilibrium, but thin enough to neglect the attenuation of the photons. The chamber shall be used to detect an air kerma rate of 1.0 mGy h^{-1}. Calculate the ionization current at this kerma rate. The mean energy of the photons is 1.0 MeV. $(\bar{W}/e)_{Ar}$=28.70 J C^{-1}.

Exercise 4.22. An ionization chamber has walls made of aluminum. The chamber is open, which means that air pressure and temperature is the same as the surrounding atmosphere. The chamber volume is 50 cm^3. The chamber wall is thick enough to give charged particle equilibrium, but thin enough to be able to neglect the attenuation of the photons. The chamber is normally used at a laboratory that is situated at sea level. However, at one occasion the chamber is used at a laboratory that is situated at an altitude of around 2000 m, resulting in a lower pressure. The chamber shall detect an air kerma rate of 3.01 Gy h^{-1}. At the first measurement, the room temperature is 295 K and the air pressure 101.5 kPa. At the second measurement at higher altitude, the room temperature is 294 K and the pressure 80.0 kPa. The mean energy of the photons is 2.0 MeV. $(\bar{W}/e)_{air}$ = 33.97 J C^{-1}. When the electrons are absorbed 0.5% of the energy is lost as photon energy. What difference in current (A) is obtained in the two measurements?

Exercise 4.23. A small ionization chamber with PMMA walls and filled with air is placed at a depth of 30 mm in a water phantom. The phantom irradiated with electrons with a mean energy at the phantom surface of 20 MeV (R_{50}=8.60 g cm^{-2}). The extrapolated range in water is 10.0 cm. The monitor of the accelerator has been calibrated to give 1.00 Gy(water)/100 monitor units at the point where the chamber is situated. After 200 monitor units a charge of 46.3 nC is obtained. How large is the effective measuring volume of the chamber? Perturbation effects due to the central electrode and scattering may be neglected. The temperature is measured to be 22°C and there is a measured pressure of 100.2 kPa. $(\bar{W}/e)_{air}$=33.97 J C^{-1}. The monitor chamber may be assumed to be independent of pressure and temperature.

4.2.5 Calorimetric and Chemical Dosimeters

Exercise 4.24. A water calorimeter is irradiated with photons with a beam quality (TPR$_{20,10}$)=0.68 and a temperature rise at the reference point of 7.50·10^{-4} °C after an irradiation time of 300 s is obtained. An NE 2581 ionization chamber is then positioned at the reference point. The dose calibration factor is $N_{D,w} = 1.758 \cdot 10^9$

Gy C^{-1} at T=273 K, p=101.3 kPa. Determine the ionization current in the chamber if T=295 K and p=100.2 kPa. The thermal effect may be neglected.

Exercise 4.25. The absorbed dose rate from a ^{60}Co-source is determined with help of a Fricke dosimeter placed in a water phantom. After 50 h irradiation an increase in absorbance at 304 nm of 0.350 is obtained. Through calibration ϵ_{304} is determined to 220 m^2 mol^{-1}. G_{Fe3+}=1.62·10^{-6} mol kg^{-1} Gy^{-1}. Calculate the absorbed dose rate to water at the point where the Fricke dosimeter is placed. The Fricke dosimeter may be considered large compared to the range of the electrons but small compared to the mean free path of the photons. ρ_{FeSO_4} = 1.024 · 10^3 kg m^{-3}. Light path length=10 mm.

Exercise 4.26. A water phantom is irradiated with electrons with the radiation quality R_{50}=20 g cm^2. At the reference depth an ionization chamber is placed. After irradiation of 60 s a charge of 3.34 nC is obtained. The temperature is 293 K and the pressure 100.2 kPa. After this a ferrous sulfate dosimeter with walls of polystyrene is placed at the measuring point. After irradiation of 100 min an absorbance of 0.832 ODU is obtained. Calculate the calibration factor k expressed in "Gy in water/nC obtained by the ionization chamber" at 295 K and 101.3 kPa.

Light path length = 10 mm. ϵG = 352·10^{-6} m^2 kg^{-1} Gy^{-1} (298 K), ρ_{FeSO_4} = 1.024 ·10^3 kg m^{-3}.

4.3 Solutions in Dosimetry

4.3.1 Definitions and Important Quantities

Solution exercise 4.1.
The energy imparted, ϵ, is defined as the sum of all energy deposits in a given volume. The mean value of energy imparted may be written as

$$\bar{\epsilon} = R_{in} - R_{out} + \sum Q \qquad (4.3.1)$$

where R_{in} is the radiant energy of all particles entering the volume. R_{out} is the radiant energy of all particles leaving the volume and $\sum Q$ is the changes in rest energy of all nuclei and elementary particles in the volume.

It is also possible to define the energy transferred as

$$\epsilon_{tr} = (R_{in})_u - (R_{out})_u^{nonr} + \sum Q \qquad (4.3.2)$$

where $(R_{in})_u$ is the radiant energy of all uncharged particles entering the volume. R_{out} is the radiant energy of all uncharged particles leaving the volume, except that energy originating from radiative losses of the kinetic energy of the charged particle.

The net energy transferred is defined as

$$\epsilon_{tr}^{n} = (R_{in})_u - (R_{out})_u^{nonr} - R_u' + \sum Q = \epsilon_{tr} - R_u' \qquad (4.3.3)$$

where R_u' is the radiant energy of all uncharged particles entering the volume.

According to Fig. 4.5 the photon entering the volume has an energy of 1.922 MeV, as 1.022 MeV is needed to produce an electron-positron pair and their total kinetic energy is 0.9 MeV. This then implies that the energy imparted is

$$\epsilon = 1.922 - 1.022 - 0.1 + (1.022 - 1.022) = 0.8\,\text{MeV}$$

The production of the electron and positron pair is compensated by the annihilation. However, the annihilation photons will have in total 0.1 MeV higher energy due to the annihilation in flight.

The energy transferred, ϵ_{tr}, is given by

$$\epsilon_{tr} = 0.7 + 0.2 = 0.9\,\text{MeV}$$

i.e. the energy transferred to the electron and the positron.

The net energy transferred, ϵ_{tr}^{n}, is given by

$$\epsilon_{tr}^{n} = 0.7 + 0.2 - 0.1 = 0.8\,\text{MeV}$$

i.e. the energy transferred to the electron and the positron reduced with the energy lost to annihilation photons due to annihilation in flight.

Solution exercise 4.2.

Curve A is obtained with the ionization chamber. Curve B with the Faraday detector and curve C with the scintillation detector. The ionization chamber measures the ionization, proportional to the absorbed dose in the chamber, which is close to the absorbed dose in water. The absorbed dose in water is nearly proportional to the electron fluence. The fluence increases first with depth due to, partly the production of secondary electrons, but mainly due to increased angular distribution of the electrons with increasing depth in the material. The fluence increases with $1/\cos\bar{\theta}$, where $\bar{\theta}$ is the mean scattering angle. The Faraday detector measures the planar fluence, which is the number of electrons per area unit independent of the angular distribution. The increase is then due to the secondary electrons only. The scintillation detector is a total absorbing detector and measures the product of the planar fluence, and the mean energy which decreases with depth. This product will decrease with depth as

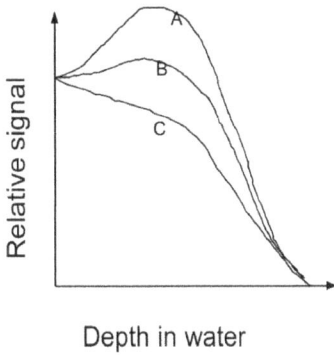

Figure 4.7: Variation of relative signal with depth for 20 MeV electrons in water obtained with A: Ionization chamber, B: Faraday cup and C: Scintillation detector.

the energy in the beam can not increase with depth. At larger depths after the dose maximum the electron fluence decreases due to loss of electrons through absorption and scattering.

4.3.2 Radiation Equilibra

Solution exercise 4.3

When charged particle equilibrium exists in a photon beam, the absorbed dose is given by

$$D = \Psi_\gamma (\mu_{en}/\rho) = \Phi_\gamma hv(\mu_{en}/\rho) \qquad (4.3.4)$$

where Φ_γ is the photon fluence, hv is the photon energy and (μ_{en}/ρ) is the mass energy absorption coefficient.

When δ-particle equilibrium is assumed to exist in an electron beam the absorbed dose is given by

$$D = \Phi_{electron}(S_{el}/\rho) \qquad (4.3.5)$$

where $\Phi_{electron}$ is the electron fluence and (S_{el}/ρ) is the mass collision stopping power.

With $\Phi_\gamma = \Phi_{electron}$ the ratio of doses is given by

$$R = \frac{(S_{el}/\rho)}{hv(\mu_{en}/\rho)} \qquad (4.3.6)$$

The energy is 1.0 MeV both for the electrons and the photons. This gives $(\mu_{en}/\rho)_{water} = 0.00310 \text{ m}^2 \text{ kg}^{-1}$ and $(S_{el}/\rho)_{water} = 0.1849 \text{ MeV m}^2 \text{ kg}^{-1}$.

Data inserted in Eq. (4.3.6) gives

$$R = \frac{0.1849}{1.0 \cdot 0.00310} = 59.6$$

Answer: The ratio between the absorbed doses in the electron and photon beam is 60.

Solution exercise 4.4

The radius of the volume, 15 mm, is large enough to assume that radiation equilibrium exists in the volume. The practical range of the β-particles from ^{32}P is around 8 mm in water. At the surface of the volume the absorbed dose is approximately half of the absorbed dose in equilibrium. The absorbed dose at the surface is obtained by integrating the dose rate over time T.

$$D = \frac{A_0 \bar{E}_\beta}{2m} \int_0^T e^{-\lambda t}\, dt = \frac{A_0 \bar{E}_\beta}{\lambda 2m}(1 - e^{-\lambda T}) \tag{4.3.7}$$

where A_0 is the initial activity and m is the mass of the volume given by

$$m = \frac{4\pi r^3 \rho}{3}$$

Data:
$\lambda = \ln 2/14.3\,\mathrm{d}^{-1}$ (half life of ^{32}P)
$T = 2.0\,\mathrm{d}$ (treatment time)
$r = 0.015\,\mathrm{m}$ (volume radius)
$\rho = 1.0 \cdot 10^3\,\mathrm{kg\,m}^{-3}$ (density of water)
$\bar{E}_\beta = 0.695\,\mathrm{MeV}$ (emitted energy/decay)
$D = 10.0\,\mathrm{Gy}$ (planned absorbed dose at surface)

Solving Eq. (4.3.7) for activity and inserting data gives

$$A_0 = \frac{10.0 \cdot 2 \cdot \ln 2 \cdot 4 \cdot \pi \cdot 0.015^3 \cdot 1.0 \cdot 10^3}{14.3 \cdot 24 \cdot 3600 \cdot 3 \cdot 0.695 \cdot 1.602 \cdot 10^{-13}(1 - e^{-2 \cdot \ln 2/14.3})} = 15.4 \cdot 10^6\,\mathrm{Bq}$$

Answer: The necessary activity is 15 MBq.

Solution exercise 4.5

The fluence of protons is not affected by the PMMA disk as the thickness is much smaller than the range of the protons. The absorbed dose to the diamond detector is, if charged particle equilibrium is assumed, given by the relation

$$D_C = \Phi(S_{\mathrm{el}}/\rho)_C \tag{4.3.8}$$

where Φ is the proton fluence and $(S_{\mathrm{el}}/\rho)_C$ is the mass collision stopping power for carbon.

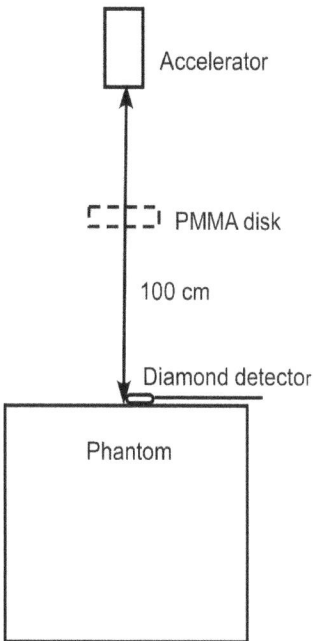

Figure 4.8: Irradiation geometry in exercise 4.5.

1. Without the PMMA-disk.

Data:

T_P=85 MeV (proton energy at accelerator surface)

$R_{85,\text{air}}$=65.48 kg m^{-2} (range of 85 MeV protons in air)

d_{air}=1.00·1.293 $\frac{273 \cdot 101}{295 \cdot 101.3}$=1.19 kg m^{-2} (air thickness with T=295 K and p=101 kPa)

The residual range after passing 100 cm of air is thus 65.48-1.19=64.3 kg m^{-2}. This corresponds to a proton energy at the diamond of 84 MeV. The stopping power in carbon of 84 MeV protons is 0.7436 MeV m^2 kg^{-1}.

2. With PMMA-disk.

The residual range after passing of 50 cm air is R=65.48-1.19/2=64.9 kg m^{-2} (cf above). This corresponds to a proton energy of 84.5 MeV.

Protons with an energy of 84.5 MeV have a range in PMMA, $R_{84.5,\text{PMMA}}$=58.72 kg m^{-2}

After 15 mm PMMA the residual range is R = 58.72−1.5·10^{-3}·1.18·10^3=41 kg m^{-2}. This corresponds to a proton energy of T=69.1 MeV. The range in air for these protons is $R_{69.1,\text{air}}$=45.2 kg m^{-2}.

After passing through 48.5 cm of air the residual range is R_{res}=45.2-0.58=44.6 kg m^{-2}. This corresponds to a proton energy of T=68.6 Mev. The stopping power in carbon of 68.6 MeV protons is 0.869 MeV m^2 kg^{-1}.

As the proton fluence is the same with and without the PMMA disk, the absorbed dose ratio is given by

$$\frac{D_{withoutPMMA}}{D_{withPMMA}} = \frac{(S/\rho)(84\,MeV)}{(S/\rho)(68.6\,MeV)} \frac{0.7436}{0.869} = 0.856$$

Answer: The signal ratio of the diamond detector without and with PMMA is 0.86.

Solution exercise 4.6

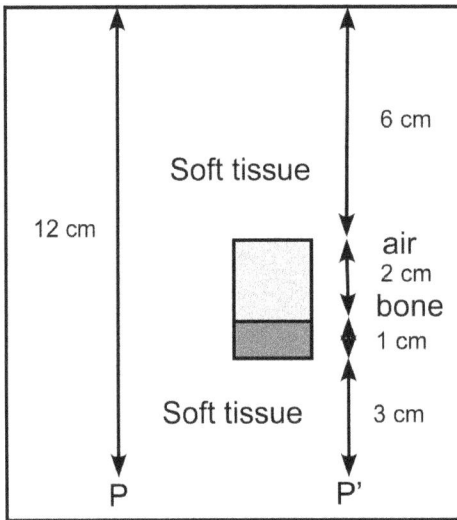

Figure 4.9: Irradiation geometry in exercise 4.6 with air and bone tissue inhomogeneity.

For protons δ-particle equilibrium is assumed to hold. The absorbed dose may then be obtained by the relation

$$D = \Phi_P(S_{el}/\rho)_{tissue} \tag{4.3.9}$$

where Φ_P is fluence of protons and $(S_{el}/\rho)_{tissue}$ is the mass collision stopping power for protons in tissue.

1. Homogeneous tissue.

The energy at the target is obtained from the following estimation:

The range of 150 MeV protons in tissue, $R_{150,\text{tissue}}$, is 159.1 kg m^{-2}. The residual range after passing a tissue thickness corresponding to 120 kg m^{-2} is 39.1 kg m^{-2}. This corresponds to an energy of 68.3 MeV. The mass collision stopping power for this energy is $(S_{\text{el}}/\rho)_{\text{tissue}}=0.964$ MeV m^2 kg^{-1}.

The planned absorbed dose at P is 2.0 Gy. The proton fluence is then

$$\Phi_P = \frac{2.0}{0.964 \cdot 1.602 \cdot 10^{-13}} \text{ m}^{-2}$$

2. Including inhomogeneous tissue.

The residual range after 6 cm of soft tissue and 2 cm of air is 99.1 kg m^{-2} assuming no energy loss in the air volume. This corresponds to an energy of 114.2 MeV.

The range of 114 MeV protons in bone tissue (density $1.8 \cdot 10^3$ kg m^{-3}) $R_{114,\text{bone}}$ is 106.0 kg m^{-2}. The residual range after 1.0 cm bone tissue is 106-18.0=88.0 kg m^{-2}. This corresponds to a proton energy of 103 MeV. In soft tissue the range of 103 MeV protons is 82.6 kg m^{-2}.

After further 3.0 cm of tissue the residual range is 82.6-30= 52.6 kg m^{-2}. This corresponds to a proton energy of 80.1 MeV. The mass collision stopping power for this energy is $(S_{\text{el}}/\rho)_{\text{tissue}}=0.852$ MeV m^2 kg^{-1}.

The absorbed dose is then given by the relation

$$D = \Phi_P(S_{\text{el}}/\rho)_{\text{tissue}} = \frac{2.0}{0.964 \cdot 1.602 \cdot 10^{-13}} 0.852 \cdot 1.602 \cdot 10^{-13} = 1.77 \text{ Gy}$$

Answer: The absorbed dose behind the inhomogeneity is 1.8 Gy.

Solution exercise 4.7
The photon fluence rate from a point source is given by

$$\dot{\Phi}_p = \frac{\sum A f_i e^{-\mu_i r}}{4\pi r^2} \tag{4.3.10}$$

where A is the activity, f_i is the yield for photon i, r is the distance from the source and μ_i is the attenuation coefficient for photon i.

This holds for the primary photons. There are also secondary photons. According to the information, the contribution from these photons is given by

$$\dot{\Phi}_s = \dot{\Phi}_p \frac{\sigma_s}{\sigma_a} \mu r \tag{4.3.11}$$

where σ_s is the Klein-Nishina coefficient for scattering and σ_a the Klein-Nishina coefficient for absorption.

For the low photon energies emitted from ^{139}Ce, charged particle equilibrium can be assumed to exist. Then the dose rate is given by

$$\dot{D} = \dot{\Phi}_p h v_p (\mu_{en}/\rho)_{water,p} + \dot{\Phi}_s h v_s (\mu_{en}/\rho)_{water,s} \qquad (4.3.12)$$

where $h v_p$ and $h v_s$ are photon energies for the primary and secondary photons respectively. $(\mu_{en}/\rho)_{water,p}$ and $(\mu_{en}/\rho)_{water,s}$ are the mass energy absorption coefficients in water for the primary and secondary photons respectively.

Using the expression for $\dot{\Phi}$ above, the relation is (only one photon energy included)

$$\dot{D} = \frac{Afe^{-\mu r}}{4\pi r^2}\left(h v_p(\mu_{en}/\rho)_{water,p} + \frac{\sigma_s}{\sigma_a}\mu r h v_s(\mu_{en}/\rho)_{water,s}\right) \qquad (4.3.13)$$

Data:
$A=0.40\cdot10^9$ Bq (source activity)
r=10 cm (distance from source)
f=0.79 (photon yield)
$h v_p$=0.166 MeV (primary photon energy. No other primary photons are included in the calculations)

The energy for the secondary photons is assumed to correspond to the mean energy of once scattered photons. Then

$h v_s$=0.134 MeV (scattered photon energy)
σ_s/σ_a=349/82.9
$(\mu_{en}/\rho)_{water,s}$=0.00270 m^2 kg^{-1} (mass energy absorption coefficient in water for primary photons)
$(\mu_{en}/\rho)_{water,p}$=0.00280 m^2 kg^{-1} (mass energy absorption coefficient in water for scattered photons)
μ = 14.5 m^{-1} (linear attenuation coefficient in water for primary photons)

Data inserted in the equation Eq. (4.3.13) gives

$$\dot{D} = \frac{0.40\cdot10^9\cdot0.79\cdot e^{-14.5\cdot0.1}}{4\pi\cdot0.1^2}\times$$

$$(0.166\cdot1.602\cdot10^{-13}\cdot0.00280 + \frac{349}{82.9}14.5\cdot0.1\cdot0.134\cdot1.602\cdot10^{-13}\cdot0.00270)$$

$\dot{D}$=0.92 mGy h^{-1}

Answer: The absorbed dose rate is 0.92 mGy h^{-1}.

Solution exercise 4.8

The absorbed dose rate to the film due to photons from the radioactive source is given by

$$\dot{D} = \frac{A \sum f_i h v_i (\mu_{en}/\rho)_{i,\text{film}}}{4\pi r^2} \qquad (4.3.14)$$

The total absorbed dose D is obtained by integrating over time t, but with a half life of 30 years the decay may be neglected. Thus the absorbed dose is obtained by just multiplying with the time, $D = \dot{D}t$.

Data:
A_0=42.5 MBq (source activity at time zero)
$f_1 = 0.898 \cdot 0.946$ (number of gamma photons per decay)
$hv_1 = 0.662$ MeV (gamma photon energy)
$(\mu_{en}/\rho)_{1,\text{film}}$=0.00306 m^2 kg^{-1} (mass energy absorption coefficient for film for 0.662 MeV)
$f_2 = 0.0605 \cdot 0.946$ (number of x rays per decay)
$hv_2 = 0.032$ MeV (x-ray mean energy)
$(\mu_{en}/\rho)_{2,\text{film}}$=1.114 m^2 kg^{-1} (mass energy absorption coefficient for film for the 0.032 MeV)
r=0.20 m (distance from source)

Data inserted in Eq. (4.3.14) gives

$$D = \frac{42.5 \cdot 10^6 \cdot 30 \cdot 3600 \cdot 1.602 \cdot 10^{-13}}{4 \cdot \pi \cdot 0.20^2}$$
$$\times (0.898 \cdot 0.946 \cdot 0.662 \cdot 0.00306 + 0.0605 \cdot 0.946 \cdot 0.032 \cdot 1.114) = 0.0055 \text{ Gy}$$

Answer: The absorbed dose to the film is 5.5 mGy.

Solution exercise 4.9

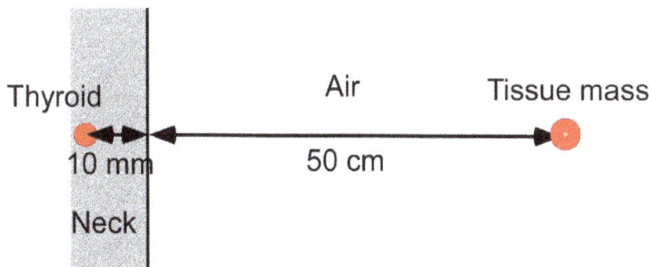

Figure 4.10: Irradiation geometry in exercise 4.9.

The total number of decays during time T is

$$\tilde{A} = \frac{A_0}{\lambda}(1 - e^{-\lambda T}) \tag{4.3.15}$$

Total energy fluence at the point of calculation will then be

$$\Psi = \frac{\tilde{A}hvf}{4\pi r^2}e^{-\mu_{\text{tissue}}d} \tag{4.3.16}$$

The total absorbed dose to a small mass of tissue at the calculation point, is assuming charged particle equilibrium, which is reasonable as the photon energy is below 500 keV, given by

$$D = \Psi(\mu_{\text{en}}/\rho)_{\text{tissue}} \tag{4.3.17}$$

Data:
$A_0=2.5\cdot10^9$ Bq (initial activity)
$\lambda=\ln2/(8.04\cdot24\cdot3600)\,\text{s}^{-1}$ (decay constant)
$T=10.0$ h (treatment time)
$hv=0.3645$ MeV (photon energy)
$f=0.812$ (number of photons per decay)
$r=51.0$ cm (distance from thyroid to calculation point)
$d=10$ mm (depth under skin)
$\mu_{\text{tissue}}=10.98\,\text{m}^{-1}$ (linear attenuation coefficient in tissue)
$(\mu_{\text{en}}/\rho)_{\text{tissue}} = 0.00322\,\text{m}^2\,\text{kg}^{-1}$ (mass energy absorption coefficient in tissue)

Combining Eq. (4.3.15), Eq. (4.3.16) and Eq. (4.3.17) and inserting data give

$$D = \frac{2.5\cdot10^9\cdot8.04\cdot24\cdot3600\cdot0.812}{\ln2\cdot4\cdot\pi\cdot0.51^2}$$
$$\times(1 - e^{-10.0\cdot\ln2/8.04\cdot24})\cdot e^{-10.98\cdot0.01}\cdot0.3645\cdot1.602\cdot10^{-13}\cdot0.00322$$

$D = 3.68\cdot10^{-3}$ Gy

Answer: The absorbed dose to tissue is 3.7 mGy.

4.3.3 Cavity Theories

Solution exercise 4.10.
The volume is large enough to give radioactive equilibrium. The absorbed dose in the milk at the center of its volume during a time T is given by

$$D_{\text{M}} = \frac{A\sum\bar{E}_{\beta,i}f_i}{\lambda m}(1 - e^{-\lambda T}) \tag{4.3.18}$$

The half life (29.12 y) is long compared to the irradiation time (7 d), thus the decrease of the activity of ^{90}Sr/^{90}Y may be neglected. The absorbed dose to the milk is instead given by

$$D_{\text{M}} = \frac{AT \sum \bar{E}_{\beta_i} f_i}{m}$$

(4.3.19)

The absorbed dose to the small LiF-dosimeter is then given by

$$D_{\text{LiF}} = s_{\text{LiF,water}} D_{\text{M}}$$

(4.3.20)

where $s_{\text{LiF,water}}$ is the mass collision stopping power ratio LiF to H$_2$O.

Solving for A gives

$$A = \frac{D_{\text{LiF}} m}{T \sum \bar{E}_{\beta_i} f_i s_{\text{LiF,water},i}}$$

(4.3.21)

Data:
A=activity of ^{90}Sr/^{90}Y
D_{LiF}= 1.85 mGy (absorbed dose to LiF dosimeter)
λ=ln2/29.12 y^{-1} (decay constant)
T=7.0 d (irradiation time)
m=100 kg (mass of milk assuming a density of $1.0 \cdot 10^3$ kg m^{-3})
$\bar{E}_{\beta,1}$=0.1957 MeV (mean energy of β-particle 1)
f_1=1.0 (relative fraction of β-particle 1)
$\bar{E}_{\beta,2}$=0.9348 MeV (mean energy of β-particle 2)
f_2=1.0 (relative fraction of β-particle 2)
$s_{\text{LiF,water},1}$(0.1957)=0.2261/0.2793 (mass collision stopping power ratio LiF to water for β-particle 1)
$s_{\text{LiF,water},2}$(0.9348)=0.1500/0.1859 (mass collision stopping power ratio LiF to water for β-particle 2)

Data inserted in Eq. (4.3.21) gives

$$A = \frac{1.85 \cdot 10^{-3} \cdot 100}{1.602 \cdot 10^{-13} \cdot 7.0 \cdot 24 \cdot 3600 \cdot [0.1957 \frac{0.2261}{0.2793} + 0.9348 \frac{0.1500}{0.1859}]}$$

A=2.092$\cdot 10^6$ Bq

Answer: The activity of ^{90}Sr/^{90}Y in the milk is 2.1 MBq

Solution exercise 4.11.

When calibrating the LiF-dosimeters with ^{60}Co the Bragg-Gray relation may be assumed to hold. Then

$$\bar{D}_{\text{LiF}} = D_{\text{water}}\bar{s}_{\text{LiF,water}} \tag{4.3.22}$$

If only Compton electrons are included in the calculations of the electron mean energy, the electron mean energy is $\bar{T}=0.584$ MeV. This is the mean energy of the emitted electrons. When passing through the phantom, the electron energy decreases and this has to be considered in order to calculate the mean energy of the electrons in equilibrium. A full calculation is rather complex, but in this context only an approximate estimation is required. The textbook by Attix (Attix, 1986) recommends that a mean value can be obtained by multiplying the initial mean energy with 0.5. This gives $\bar{\bar{T}}=0.5\cdot0.584=0.292$ MeV.

The calibration constant R_{LiF}, expressed in pulses per Gy_{LiF}, is given by the relation

$$R_{\text{LiF}} = \frac{R_{\text{water}}(S_{\text{el}}/\rho)_{\text{water}}}{(S_{\text{el}}/\rho)_{\text{LiF}}} \tag{4.3.23}$$

Data:
R_{water}=37200 (pulses per Gy in water)
$(S_{\text{el}}/\rho)_{\text{LiF}}$=0.1926 MeV m^2 kg^{-1} (mass collision stopping power for LiF)
$(S_{\text{el}}/\rho)_{\text{water}}$=0.2378 MeV m^2 kg^{-1} (mass collision stopping power for water)

Data inserted in Eq. (4.3.23) gives

$$R_{\text{LiF}} = \frac{37200\cdot0.2378}{0.1926} = 45930 \text{ pulses per Gy in LiF}$$

When irradiating the LiF dosimeters with $^{99\text{m}}$Tc, a charged particle equilibrium is assumed. Secondary photons are also neglected. The absorbed dose to the polystyrene container is given by

$$D_{\text{C}_8\text{H}_8} = D_{\text{LiF}}\frac{(\mu_{\text{en}}/\rho)_{\text{C}_8\text{H}_8}}{(\mu_{\text{en}}/\rho)_{\text{LiF}}} \tag{4.3.24}$$

The absorbed dose to water is given by

$$D_{\text{water}} = \bar{D}_{\text{C}_8\text{H}_8}\frac{(\mu_{\text{en}}/\rho)_{\text{water}}}{(\mu_{\text{en}}/\rho)_{\text{C}_8\text{H}_8}} = \bar{D}_{\text{LiF}}\frac{(\mu_{\text{en}}/\rho)_{\text{water}}}{(\mu_{\text{en}}/\rho)_{\text{LiF}}} = \frac{M_{\text{LiF}}(\mu_{\text{en}}/\rho)_{\text{water}}}{R_{\text{LiF}}(\mu_{\text{en}}/\rho)_{\text{LiF}}} \tag{4.3.25}$$

Data:
$h\nu$=140 keV (x rays are neglected)
$(\mu_{\text{en}}/\rho)_{\text{water}}$=2.700$\cdot10^{-3}$ m^2 kg^{-1} (mass energy absorption coefficient for water)
$(\mu_{\text{en}}/\rho)_{\text{LiF}}$=2.308$\cdot10^{-3}$ m^2 kg^{-1} (mass energy absorption coefficient for LiF)
D_{water}=0.10 Gy
M_{LiF}=number of pulses when irradiating with $^{99\text{m}}Tc$

Data inserted in Eq. (4.3.25) gives

$$0.10 = \frac{M_{\text{LiF}}\cdot2.700\cdot10^{-3}}{45930\cdot2.308\cdot10^{-3}}$$

M_{LiF}=3926 pulses

Answer: $3.93 \cdot 10^3$ pulses are needed for an absorbed dose to water of 0.10 Gy.

Solution exercise 4.12.
The absorbed dose to the dosimeter is given by

$$\bar{D}_{dos,large} = D_{water}\frac{(\mu_{en}/\rho)_{dos}}{(\mu_{en}/\rho)_{water}} \text{ large detector} \tag{4.3.26}$$

or

$$\bar{D}_{dos,small} = D_{water}\frac{(S_{el}/\rho)_{dos}}{(S_{el}/\rho)_{water}} \text{ small detector} \tag{4.3.27}$$

Large detector

Data:
D_w=2.00 Gy (absorbed dose to water)
hv=4.0 MeV (photon energy)
$(\mu_{en}/\rho)_{Si}$=0.001963 m^2 kg^{-1} (mass energy absorption coefficient for Si)
$(\mu_{en}/\rho)_{water}$=0.00206 m^2 kg^{-1} (mass energy absorption coefficient for water)
$(\mu_{en}/\rho)_C$=0.00185 m^2 kg^{-1} (mass energy absorption coefficient for C. Diamond is considered as carbon)

Data inserted in Eq. (4.3.26) gives

$$\bar{D}_{Si} = 2.0\frac{0.001963}{0.00206} = 1.91 \text{ Gy}$$

and

$$\bar{D}_C = 2.0\frac{0.00185}{0.00206} = 1.80 \text{ Gy}$$

Small detector

For determining the mean energy of the electrons produced in photon interaction, both Compton scattering and pair production in the field of the nucleus are included.

The mean energy is obtained through the relation

$$\bar{T} = \frac{0.00322 \cdot 2.428}{0.00358} + \frac{0.000181 \cdot 1.50 \cdot 2}{0.00358} = 2.34 \text{ MeV}$$

This is the mean energy of the emitted electrons. The mean energy of the electrons in

Table 4.1: Data for calculation of the mean energy of electrons in water.

Type of interaction	$\sigma(m^2\,kg^{-1})$	number of electrons	T/MeV
Compton	0.00322	1	2.428
pair production	0.000181	2	1.50
$\sum \sigma_i n_i$:	0.00358		2.34

equilibrium, using the factor 0.5, gives a mean energy of 1.17 MeV.

Data:
$(S_{el}/\rho)_{Si}$ = 0.1501 MeV m^2 kg^{-1} (mass collision stopping power for Si)
$(S_{el}/\rho)_{C}$ = 0.1603 MeV m^2 kg^{-1} (mass collision stopping power for C)
$(S_{el}/\rho)_{water}$=0.1833 MeV m^2 kg^{-1} (mass collision stopping power for water)

Data inserted in Eq. (4.3.27) gives

$$\bar{D}_{Si} = 2.0\frac{0.1501}{0.1833} = 1.64\text{ Gy}$$

and

$$\bar{D}_{C} = 2.0\frac{0.1603}{0.1833} = 1.75\text{ Gy}$$

Answer: The absorbed doses to the diamond detectors are 1.80 Gy (large detector) and 1.75 Gy (small detector). The absorbed doses to the Si-detectors are 1.91 Gy (large detector) and 1.64 Gy (small detector).

Solution exercise 4.13.
1. Water phantom.

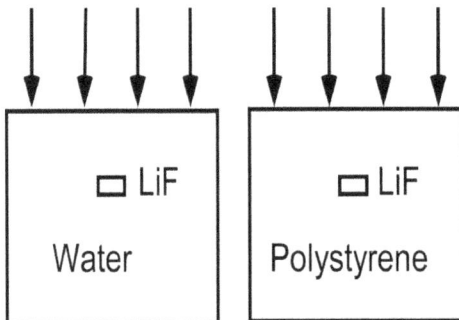

Figure 4.11: Irradiation geometry in exercise 4.13.

If the LiF dosimeter is small compared to the range of the electrons and the Bragg-Gray relation is assumed to hold, the absorbed dose to the LiF-dosimeter is given by

$$\bar{D}_{LiF} = D_{water}\bar{s}_{LiF,water} \qquad (4.3.28)$$

where $\bar{s}_{LiF,water}$ is the mean collision stopping power ratio LiF to water.

As there is charged particle equilibrium the absorbed dose to water can be obtained from

$$D_{water} = \Phi h\nu(\mu_{en}/\rho)_{water} \qquad (4.3.29)$$

where Φ is the photon fluence, $h\nu$ is the photon energy and $(\mu_{en}/\rho)_{water}$ is the mass energy absorption coefficient for water.

Combining the equations Eq. (4.3.28) and Eq. (4.3.29) the absorbed dose to LiF is given by

$$\bar{D}_{LiF} = \Phi h\nu(\mu_{en}/\rho)_{water}\bar{s}_{LiF,water} \qquad (4.3.30)$$

2. Polystyrene phantom.

The absorbed dose to the LiF dosimeter is given by

$$\bar{D}_{LiF} = \Phi h\nu(\mu_{en}/\rho)_{C_8H_8}\bar{s}_{LiF,C_8H_8} \qquad (4.3.31)$$

Combining Eq. 4.3.30 and Eq. 4.3.31 and assuming the same fluence and energy distribution gives

$$\frac{\bar{D}_{LiF,water}}{\bar{D}_{LiF,C_8H_8}} = \frac{\Phi h\nu(\mu_{en}/\rho)_{water}\bar{s}_{LiF,water}}{\Phi h\nu}(\mu_{en}/\rho)_{C_8H_8}\bar{s}_{LiF,C_8H_8} \qquad (4.3.32)$$

The mean photon energy is assumed to be 2.5 MeV.

The stopping power values are taken for the electron mean energy at LiF in each phantom.

1. Water phantom.

The mean energy is obtained through the relation

Table 4.2: Data for calculation of the mean energy of electrons in water.

Type of interaction	$\sigma(m^2\,kg^{-1})$	number of electrons	T/MeV
compton	0.004375	1	1.3845
pair production	0.0000756	2	0.75
$\sum \sigma_i n_i$:	0.004526		1.36

$$\bar{T} = \frac{0.004375 \cdot 1.3845}{0.004526} + \frac{0.0000756 \cdot 0.75 \cdot 2}{0.004526} = 1.36\,\text{MeV}$$

This is then the mean energy of the emitted electrons. The mean energy of the electrons in equilibrium, using the factor 0.5, gives the mean energy 0.68 MeV.

2. Polystyrene phantom.

Table 4.3: Data for calculation of the mean energy of electrons in polystyrene.

Type of interaction	$\sigma(\text{m}^2\,\text{kg}^{-1})$	number of electrons	T/MeV
Compton	0.000424	1	1.3845
pair production	0.0000584	2	0.75
$\sum \sigma_i n_i$:	0.004357		1.37

The results show that there is a very small difference in mean energy between the two phantom materials and the same value 0.68 MeV is used for both phantoms. Thus the values of $(S_{el}/\rho)_{LiF}$ cancel in Eq. (4.3.32).

Data:
$(\mu_{en}/\rho)_{water}$=0.00244 m^2 kg^{-1} (mass energy absorption coefficient for water)
$(\mu_{en}/\rho)_{C_8H_8}$=0.00236 m^2 kg^{-1} (mass energy absorption coefficient for polystyrene)
$(S_{el}/\rho)_{water}$=0.19245 MeV m^2 kg^{-1} (mass collision stopping power for water)
$(S_{el}/\rho)_{C_8H_8}$=0.1872 MeV m^2 kg^{-1} (mass collision stopping power for polystyrene)

Data inserted in Eq. (4.3.32) gives

$$\frac{\bar{D}_{LiF,water}}{\bar{D}_{LiF,C_8H_8}} = \frac{0.00244 \cdot 0.1872}{0.00236 \cdot 0.19245} = 1.006$$

Answer: The ratio of the absorbed dose to the LiF dosimeter in water to the LiF dosimeter in polystyrene is 1.01.

Solution exercise 4.14.
The absorbed dose to the LiF dosimeter is according to the Burlin theory

$$\bar{D}_{LiF} = D_{water}(d\bar{s}_{LiF,water} + (1-d)(\bar{\mu}_{en}/\rho)_{LiF,water}) \qquad (4.3.33)$$

where the weighting factor, d, according to Burlin is given by the relation

$$d = \frac{1 - e^{-\beta L}}{\beta L} \qquad (4.3.34)$$

where β is the absorption coefficient in the dosimeter for the secondary electrons emitted by the photons. L is the mean chord length in the dosimeter. For a convex volume, L is given by

$$L = 4V/S \qquad (4.3.35)$$

where V and S are the dosimeter volume and surface area respectively.

$\bar{s}_{\text{LiF,water}}$ shall be determined for the mean energy of the emitted electrons. The mean energy of the electrons is obtained by assuming that the photons interact through Compton interaction and pair production in the nuclear field, as the mean photon energy is 4 MeV. This is then the mean energy of the emitted electrons. The mean energy of the electrons in equilibrium, using the factor 0.5, gives the mean energy 1.17 MeV.

Table 4.4: Data for calculation of the mean energy of electrons in water.

Type of interaction	$\sigma(\text{m}^2 \text{ kg}^{-1})$	number of electrons	T/MeV
Compton	0.00322	1	2.428
pair production	0.000186	2	1.50
$\sum \sigma_i n_i$:	0.00359		2.33

The LiF dosimeter has a cylindrical shape and then L is given by

$$L = \frac{4\pi r^2 h}{2\pi r^2 + 2\pi rh} \qquad (4.3.36)$$

Data:
r=6.5 mm (radius of the dosimeter)
h=0.5 mm (height of the dosimeter)
$\bar{s}_{\text{LiF,water}}$=1.481/1.840=0.805 (mass collision stopping power ratio, LiF to water).
$(\bar{\mu}_{en}/\rho)_{\text{LiF,water}}$=0.00173/0.00206=0.840 (mass energy absorption coefficient ratio LiF to water)
$\bar{D}_{\text{LiF}}$=1.47 Gy

Data inserted in Eq. (4.3.36) gives

$$L = \frac{4\pi \cdot 13^2 \cdot 0.5}{2\pi 6.5^2 + 2\pi \cdot 6.5 \cdot 0.5} = 1.73 \text{ mm}$$

The density of LiF is 2.30·10^3 kg m^{-3} and thus L=3.98 kg m^{-2}.

The absorption coefficient is obtained using the relation

$$\beta = \frac{3.5Z}{AT_{\max}^{1.14}} \text{ m}^2 \text{ kg}^{-1}. \qquad (4.3.37)$$

Assuming that the maximum electron energy, T_{max} is the energy of the most energetic Compton electron gives $T_{max}=3.75$ MeV if the photon energy is 4 MeV. Setting Z, atomic number, and A, mass number, for LiF to 12 and 26 respectively Eq. (4.3.37) gives d

$$\beta = \frac{3.5 \cdot 12}{26 \cdot 3.75^{1.14}} = 0.358 \, \text{m}^2 \, \text{kg}^{-1}$$

Inserting the values of L and β in Eq. (4.3.34) give

$$d = \frac{1 - e^{0.358 \cdot 3.98}}{0.358 \cdot 3.98} = 0.533 \tag{4.3.38}$$

This and data for $\bar{s}_{\text{LiF,water}}$ and $(\bar{\mu}_{en}/\rho)_{\text{LiF,water}}$ inserted in Eq. (4.3.33) gives

$$1.47 = D_{\text{water}}(0.533 \cdot 0.840 + (1 - 0.533)0.8398)$$

D_{water}=1.750 Gy

Answer: The absorbed dose to water is 1.75 Gy.

Solution exercise 4.15.
Case 1: Primary photons; the detector is assumed to be small compared to the range of electrons. The Bragg-Gray relation is assumed to hold. Then

$$\bar{D}_{\text{Si}} = D_{\text{water}}\bar{s}_{\text{Si,water}} \tag{4.3.39}$$

The mean energy of the electrons is obtained by assuming that the photons interact through Compton interaction and pair production in the nuclear field. The mean photon energy is 6 MeV.

Table 4.5: Data for calculation of the mean energy of electrons in water.

Type of interaction	$\bar{\sigma}(\text{m}^2 \, \text{kg}^{-1})$	number of electrons	T/MeV
Compton	0.00245	1	3.864
pair production	0.000299	2	2.5
$\sum \sigma_i n_i$:	0.003048		3.60

The mean energy is obtained through the relation

$$\bar{T} = \frac{0.002425 \cdot 3.864}{0.003048} + \frac{0.000299 \cdot 2.5 \cdot 2}{0.003048} = 3.60 \, \text{MeV}$$

This is the mean energy of the emitted electrons. The mean energy of the electrons in equilibrium, using the factor 0.5, gives the mean energy 1.80 MeV.

Data:

$(S_{el}/\rho)_{Si}$=0.1511 MeV m^2 kg^{-1} (mass collision stopping power for Si)
$(S_{el}/\rho)_{water}$=0.1821 MeV m^2 kg^{-1} (mass collision stopping power for water)

Data inserted gives

$$\bar{D}_{Si} = D_{water}\frac{0.1511}{0.1821} = 0.83 D_{water}$$

Case 2: Secondary photons; the detector is assumed to be large compared to the range of the electrons. Then the absorbed dose ratio is given by the ratio of the mass energy absorption coefficients.

$$\bar{D}_{Si} = D_{water}(\mu_{en}/\rho)_{Si,water} \qquad (4.3.40)$$

Data:

$(\mu_{en}/\rho)_{Si}$=0.00291 m^2 kg^{-1} (mass energy absorption coefficient for Si)
$(\mu_{en}/\rho)_{water}$=0.00297 m^2 kg^{-1} (mass energy absorption coefficient for water)

Data inserted gives

$$D_{Si} = D_{water}\frac{0.00291}{0.00297} = 0.98 D_{water}$$

Answer: The absorbed dose ratio $(\bar{D}_{Si}/D_{water})$ is 0.82 for high energy photons and 0.98 for low energy photons.

Solution 4.16

The absorbed dose to the diamond detector without the polystyrene cover is given by

$$\bar{D}_{C,1} = D_{water}\bar{s}_{C,water} \qquad (4.3.41)$$

The absorbed dose to the diamond detector with the polystyrene cover is given by

$$D_{C,2} = D_{water}(\mu_{en}/\rho)_{C_8H_8,water}\bar{s}_{C,C_8H_8} \qquad (4.3.42)$$

where $\bar{s}_{C,C_8H_8}$ is the mass collision stopping power ratio carbon to polystyrene and $(\mu_{en}/\rho)_{C_8H_8,water}$ is the mass energy absorption coefficient ratio polystyrene to water. The stopping power ratio is calculated for the mean energy of the electrons.

In equilibrium the mean energy is 0.5·2.97=1.48 MeV. Assume the same mean energy in water and in the polystyrene cover.

Data:

Table 4.6: Data for calculation of the mean energy of electrons in water.

Type of interaction	$\sigma(m^2\,kg^{-1})$	number of electrons	T/MeV
Compton	0.00278	1	3.14
pair production	0.000243	2	2
$\sum \sigma_i n_i$:	0.00327		2.97 MeV

$(S_{el}/\rho)_C$=0.1593 MeV m^2 kg^{-1} (mass collision stopping power for carbon)
$(S_{el}/\rho)_{C_8H_8}$=0.1766 MeV m^2 kg^{-1} (mass collision stopping power for polystyrene)
$(S_{el}/\rho)_{water}$=0.1823 MeV m^2 kg^{-1} (mass collision stopping power for water)
$(\mu_{en}/\rho)_{C_8H_8}$=0.00182 m^2 kg^{-1} (mass energy absorption coefficient for polystyrene)
$(\mu_{en}/\rho)_{water}$=0.00191 m^2 kg^{-1} (mass energy absorption coefficient for water)

Data inserted in Eq. (4.3.41) and Eq. (4.3.42) gives

$$\bar{D}_{C,1} = D_{water}\frac{0.1593}{0.1823} = 0.874 D_{water}$$

and

$$\bar{D}_{C,2} = D_{water}\frac{0.00182}{0.00191}\frac{0.1593}{0.1823} = 0.860 D_{water}$$

The ratio of absorbed doses with and without the polystyrene cover is 1.017.

Answer: The ratio of the absorbed doses in the diamond detector with and without the polystyrene cover is 1.017.

Solution exercise 4.17.
When irradiated with 40 MV x rays (13 MeV) the LiF-dosimeters can be regarded as small compared to the range of the electrons and the Bragg-Gray relation may be assumed to hold.

Calibration in the Teflon phantom, assuming Teflon is equivalent with LiF, results in a charged particle equilibrium in LiF. Then

$$\bar{D}_{LiF} = K_{air}(\mu_{en}/\rho)_{LiF,air}k_{att} \tag{4.3.43}$$

The calibration constant, R is given by

$$R = \bar{D}_{LiF}/M_{cal} = \dot{K}_{cal,air}t_{cal}(\mu_{en}/\rho)_{LiF,air}k_{att}/M_{cal} \tag{4.3.44}$$

Irradiation in the polystyrene phantom gives

$$D_{C_8H_8} = \bar{D}_{LiF}\bar{s}_{C_8H_8,LiF} = RM_{C_8H_8}\bar{s}_{C_8H_8,LiF} \tag{4.3.45}$$

The absorbed dose rate to polystyrene is then given by

$$\dot{D}_{C_8H_8} = M_{C_8H_8}\bar{s}_{C_8H_8,LiF}\dot{K}_{cal,air}t_{cal}(\mu_{en}/\rho)_{LiF,air}k_{att}/(M_{cal}t_{C_8H_8}) \tag{4.3.46}$$

Data:

$\dot{K}_{cal,air}$=0.63 Gy min^{-1} (collision kerma rate in air)

t_{cal}=120 s (irradiation time at calibration)

M_{cal}= 37500 (number of pulses at calibration)

$(\mu_{en}/\rho)_{LiF}$=0.00247 m^2 kg^{-1} (mass energy absorption coefficient for LiF)

$(\mu_{en}/\rho)_{air}$=0.00267 m^2 kg^{-1} (mass energy absorption coefficient for air)

$M_{C_8H_8}$= 42300 (number of pulses at irradiation in polystyrene phantom)

$t_{C_8H_8}$=180 s (irradiation time at irradiation in polystyrene phantom)

k_{att}=0.99 (correction for attenuation in the Teflon wall)

The stopping power ratio is calculated for the mean energy of the electrons.

Table 4.7: Data for calculation of the mean energy of electrons in polystyrene.

Type of interaction	σ(m^2 kg^{-1})	number of electrons	T/MeV
Compton	0.00140	1	9.1
pair production, nuclear field	0.000433	2	6.0
pair production, electron field	0.000051	3	4.0
$\sum \sigma_i n_i$:	0.002419		7.67 MeV

In equilibrium the mean electron energy is $\bar{T}$=0.5·7.67=3.83 MeV. This gives

$(S_{el}/\rho)_{C_8H_8}$=0.1812 MeV m^2 kg^{-1} (mass collision stopping power for polystyrene)

$(S_{el}\rho)_{LiF}$=0.1510 MeV m^2 kg^{-1} (mass collision stopping power for LiF)

Data inserted in Eq. (4.3.46) gives

$$\dot{D}_{C_8H_8} = \frac{42300 \cdot 0.63 \cdot 2.0 \cdot 0.99 \cdot 0.1812 \cdot 0.00247}{0.151 \cdot 0.00267 \cdot 37500 \cdot 180}$$

$\dot{D}_{C_8H_8}$=0.00868 Gy s^{-1}

Answer: The absorbed dose rate to polystyrene at the measuring point is 8.7 mGy s^{-1}.

Solution exercise 4.18.

Assuming that the perturbation effects do not change with depth, the ratios of measured values at the depth of z_{max} and at z_{50} are given by the relation

$$\frac{M(z_{50})}{M(z_{max})} = 0.5 \frac{\bar{s}(z_{max})_{water,det}}{\bar{s}(z)_{water,det}} \qquad (4.3.47)$$

where $\bar{s}(z_{50})_{\text{water,det}}$ is the mass collision stopping power ratio water to detector at depth z_{50} and $\bar{s}(z_{\text{max}})_{\text{water,det}}$ is the mass collision stopping power ratio water to detector at depth z_{max}.

The stopping power ratios $s_{\text{water,air}}$ may be calculated using Table 20 IAEA TRS 398 (IAEA 2000) using the information that $R_{50}=80$ kg m^{-2}.

For the Si-diode the stopping power ratios are obtained by estimating the mean electron energy at different depths using the relation proposed by Brahme (Brahme, 1975).

$$\bar{T}(z) = \bar{T}_0 - (S/\rho)_{\text{water,tot,0}} \frac{1 - e^{-\rho z \epsilon_{\text{water,rad,0}}}}{\epsilon_{\text{water,rad,0}}} \qquad (4.3.48)$$

where

$$\epsilon_{\text{water,rad,0}} = \frac{1}{\bar{T}_0}(S/\rho)_{\text{water,rad}} \qquad (4.3.49)$$

1. Ionization chamber

$\bar{s}(30, z_{\text{max}})_{\text{water,air}}=0.9975$ (mass collision stopping power ratio water to air at depth 30 kg m^{-2})
$\bar{s}(80, z_{50})_{\text{water,air}}=1.083$ (mass collision stopping power ratio water to air at depth 80 kg m^{-2})

2. Si-detector

$(S/\rho)_{\text{water,tot,0}}=0.24546$ MeV kg m^{-2} (total mass stopping power for the mean electron energy at surface, 20 MeV)
$\epsilon_{\text{water,rad,0}}=0.04086/20=0.002043$ kg/m^{-2} (radiation mass stopping power divided with the mean electron energy at surface, 20 MeV).

Data inserted in Eq. (4.3.48) for $z_{\text{max}}=30$ kg m^{-2} gives

.

$$\bar{T}(30) = 20 - 0.24546 \frac{1 - e^{-30 \cdot 0.002043}}{0.002043} = 12.9 \text{ MeV}$$

The corresponding energy at depth $z=80$ kg m^{-2} is 1.9 MeV

The mass collision stopping power ratios water to Si are then

$\bar{s}(30, z_{\text{max}})_{\text{water,Si}}=0.1997/0.1742=1.146$
$\bar{s}(80, z_{50})_{\text{water,Si}}=0.1822/0.1515=1.203$

Data inserted in Eq. (4.3.47) gives the ratios as

1. Ionization chamber

$$\frac{M(z_{50})}{M(z_{\max})} = 0.5\frac{0.9975}{1.083} = 0.460$$

2. Si-detector

$$\frac{M(z_{50})}{M(z_{\max})} = 0.5\frac{1.146}{1.203} = 0.476$$

Answer: The ratios of the measured values at z_{50} are 0.46 for the ionization chamber and 0.48 for the Si-detector.

4.3.4 Ionization Chamber Dosimetry

Solution exercise 4.19.

The absorbed dose to water is given by the equation

$$D_{\text{water},Q} = M_Q N_{D,w} k_Q \qquad (4.3.50)$$

where M_Q is the measured value corrected for influence quantities

$$M_Q = M_R k_{Tp} k_s \qquad (4.3.51)$$

Data:

$N_{D,w}$=0.0139 Gy nC^{-1} (calibration factor for the NE2571 chamber calibrated in ^{60}Co at T=293 K and p=101.3 kPa)

k_Q=0.998 (chamber specific factor to correct for the difference between the reference quality and the specific quality used, Q. Table 14 (IAEA, 2000))

M_R=213.3 nC (measured charge)

Correction factor k_{Tp} for temperature and pressure is given by

$$k_{Tp} = \frac{101.3 \cdot 295}{100 \cdot 293}$$

The correction factor for recombination k_s is given by (IAEA TRS 398)

$$k_s = a_0 + a_1(\frac{M_1}{M_2}) + a_2(\frac{M_1}{M_2})^2 \qquad (4.3.52)$$

where M_1/M_2=1.012 if V_1=300 V and V_2=100 V. Then the constants are (Table 9 IAEA TRS 398) $a_0 = 1.198$, $a_1 = -0.875$, $a_2 = 0.677$. Data inserted in Eq. (4.3.52) gives

$$k_s = 1.198 - 0.875 \cdot 1.012 + 0.677 \cdot 1.012^2 = 1.00585$$

Data inserted in Eq. (4.3.50) gives

$$D_{\text{water},Q} = 213.3 \cdot 1.00585 \cdot 0.998 \cdot 0.0139 \frac{101.3 \cdot 295}{100 \cdot 293} = 3.036\,\text{Gy}$$

The monitor value is 3.00 Gy. This gives the calibration factor equal to 3.036/3.00=1.012.

Answer: The calibration factor is 1.012.

Solution exercise 4.20.
The absorbed dose rate to water is given by

$$\dot{D}_{\text{water},Q} = M_Q N_{D,w} k_Q / t \tag{4.3.53}$$

M_Q is the measured value corrected for influence quantities

$$M_Q = M_R k_{Tp} k_s \tag{4.3.54}$$

Data:
$N_{D,w}$=9.17·10^{-3} Gy nC^{-1} (calibration factor for the NE2571 chamber calibrated in ^{60}Co, at T=293 K and p=101.3 kPa)
k_Q=0.904 (chamber specific factor to correct for the difference between the reference quality and the specific quality used, Q. Table 18 (IAEA, 2000))
M_R=235 nC (measured charge, T=295 K, p=98.3 kPa)
t = 120 s (irradiation time)
   Correction factor k_{Tp} for temperature and pressure is given by

$$k_{Tp} = \frac{101.3 \cdot 295}{98.3 \cdot 293}$$

The correction factor for recombination k_s is given by (IAEA, 2000)

$$k_s = a_0 + a_1\left(\frac{M_1}{M_2}\right) + a_2\left(\frac{M_1}{M_2}\right)^2 \tag{4.3.55}$$

where M_1/M_2=1.010 if V_1=400 V and V_2=100 V. Then the constants are (Table 9 (IAEA, 2000)) a_0 = 1.022, a_1 = −0.363, a_2 = 0.341. Data inserted gives

$$k_s = 1.022 - 0.363 \cdot 1.010 + 0.341 \cdot 1.010^2 = 1.0032$$

Data inserted in Eq. (4.3.53) gives

$$\dot{D}_{\text{water},Q} = 9.17 \cdot 10^{-3} \cdot 1.0032 \cdot 0.904 \cdot 235 \frac{101.3 \cdot 295}{98.3 \cdot 293} \frac{1}{120} = 0.0168\,\text{Gy s}^{-1}$$

Answer: The absorbed dose rate is 17 mGy s^{-1}.

Solution exercise 4.21.

Assume that the chamber is a Bragg-Gray cavity, as the range of the electrons is much larger than the size of the chamber. Then

$$\bar{D}_{Ar} = D_{Al}\bar{s}_{Ar,Al} \tag{4.3.56}$$

The chamber wall of aluminum is thick compared to the range of the electrons. Thus there is charged particle equilibrium in the wall. Then

$$D_{Al} = \Psi_{air}(\mu_{en}/\rho)_{Al} \tag{4.3.57}$$

where Ψ_{air} is the photon energy fluence in air at the chamber. The kerma to air is given by

$$K_{air} = \Psi_{air}(\mu_{tr}/\rho)_{air} \tag{4.3.58}$$

Thus the dose to the aluminum wall can be expressed as

$$D_{Al} = K_{air}\frac{(\mu_{en}/\rho)_{Al}}{(\mu_{tr}/\rho)_{air}} \tag{4.3.59}$$

and

$$\bar{D}_{Ar} = K_{air}\frac{(\mu_{en}/\rho)_{Al}}{(\mu_{tr}/\rho)_{air}}\bar{s}_{Ar,Al} \tag{4.3.60}$$

The current from the ionization chamber is given by

$$I = \frac{\dot{D}_{Ar}m_{Ar}}{(\bar{W}/e)_{Ar}} \tag{4.3.61}$$

Data:

hv=1.0 Mev (photon energy)

$(\mu_{en}/\rho)_{Al}$=0.00268 m^2 kg^{-1} (mass energy absorption coefficient for Al)

$(\mu_{tr}/\rho)_{air}$=0.00279 m^2 kg^{-1} (mass energy transfer coefficient for air)

K_{air}=1.0 mGy h^{-1}

The mean electron energy is obtained by assuming that only Compton interaction is of importance and that the mean energy in equilibrium is half of the emitted energy.

$$\bar{T} = 0.5 \cdot 0.44 = 0.22\,\text{MeV}$$

$(S_{el}/\rho)_{Ar}=0.1920\,\text{MeV m}^2\,\text{kg}^{-1}$ (mass collision stopping power for Ar)
$(S_{el}/\rho)_{Al}=0.2082\,\text{MeV m}^2\,\text{kg}^{-1}$ (mass collision stopping power for air)
$(\bar{W}/e)_{Ar}=28.70\,\text{J C}^{-1}$ (mean energy expended to produce a charge of 1 C in Ar)

The mass of the chamber volume is obtained through the relation

$$m = \rho_{Ar} v \tag{4.3.62}$$

The density of argon is given by

$$\rho(T, p) = \rho(NTP)\frac{T_0 p}{T p_0} \tag{4.3.63}$$

Data:
$\rho(NTP)=1.784\,\text{kg m}^{-3}$ (density of Ar at NTP)
$T=293\,\text{K}$, $p=800\,\text{kPa}$, $T_0=273.2\,\text{K}$, $p_0=101.3\,\text{kPa}$
$v=50\cdot10^{-6}\,\text{m}^3$ (chamber volume)

Data inserted in Eq. (4.3.61) gives

$$I = \frac{1.0 \cdot 10^{-3} \cdot 0.00268 \cdot 0.1920 \cdot 1.784 \cdot 800 \cdot 273.2 \cdot 50 \cdot 10^{-6}}{0.00279 \cdot 0.2082 \cdot 28.70 \cdot 101.3 \cdot 293 \cdot 3600} = 5.63 \cdot 10^{-12}\,\text{A}$$

Answer: The ionization current is 5.63 pA.

Solution exercise 4.22.
Assuming charged particle equilibrium in aluminum and that the Bragg-Gray relation holds for the air cavity, the current in the ionization chamber is given by

$$I = \dot{D}_{air} m/(\bar{W}/e)_{air} \tag{4.3.64}$$

where

$$\dot{D}_{air} = \dot{K}_{air}(\mu_{en}/\rho)_{Al,air}\bar{s}_{air,Al}(1 - g) \tag{4.3.65}$$

This implies that the current is proportional to the mass and then the difference in current is given by

$$\Delta I = \dot{D}_{air}\Delta m/(\bar{W}/e)_{air} \tag{4.3.66}$$

The mean energy of the electrons is obtained by assuming that only Compton interaction is important. With a photon energy of 2.0 MeV, the mean Compton electron energy is 1.06 MeV. The electron energy in equilibrium is therefore $0.5\cdot1.06\,\text{MeV}=0.53\,\text{MeV}$.

Data:
$\dot{K}_{air}=3.01\,\text{Gy h}^{-1}$ (kerma rate in air)

$(\mu_{en}/\rho)_{Al}$=0.00226 m^2 kg^{-1} (mass energy absorption coefficient for Al)
$(\mu_{en}/\rho)_{air}$=0.00234 m^2 kg^{-1} (mass energy absorption coefficient for air)
$(S_{el}/\rho)_{air}$=0.1874 MeV m^2 kg^{-1} (mass collision stopping power for air)
$(S_{el}/\rho)_{Al}$=0.1655 MeV m^2 kg^{-1} (mass collision stopping power for Al)
$(\bar{W}/e)_{air}$=33.97 J C^{-1} (mean energy expended to produce a charge of 1 C in air)
$(1\text{-}g)$=0.995 (fraction of energy used for nonradiative collision losses)
Measurement 1: T=295 K, p=101.5 kPa
Measurement 2: T=294 K, p=80.0 kPa

The air mass of the ionization chamber is

$$m_1 = 1.293 \cdot 50 \cdot 10^{-6} \frac{273 \cdot 101.5}{295 \cdot 101.3} = 5.995 \cdot 10^{-5} \text{ kg}$$

and

$$m_2 = 1.293 \cdot 50 \cdot 10^{-6} \frac{273 \cdot 80.0}{294 \cdot 101.3} = 4.741 \cdot 10^{-5} \text{ kg}$$

This gives the difference in mass, Δm=1.254·10^{-5} kg

Then the difference in current is

$$\Delta I = \frac{3.01 \cdot 0.00226 \cdot 0.1874 \cdot 1.254 \cdot 10^{-5} \cdot 0.995}{3600 \cdot 0.00234 \cdot 0.1655 \cdot 33.97} = 3.36 \cdot 10^{-10} \text{ A}$$

Answer: The difference in ionization current is 0.34 nA.

Solution exercise 4.23.

The reference depth for the radiation quality R_{50}=8.60 g cm^{-2} is given by the relation (Table 17 (IAEA, 2000))

$$z_{ref} = 0.6R_{50} - 0.1 = 0.6 \cdot 8.60 - 0.1 = 5.06 \text{ g cm}^{-2} \qquad (4.3.67)$$

The energy at the reference depth is obtained from the relation (IAEA, 2000)

$$E_{z,ref} = 0.07 + 1.027R_{50} - 0.0048(R_{50})^2$$
$$= 0.07 + 1.027 \cdot 8.60 - 0.0048 \cdot 8.60^2 = 8.55 \text{ MeV}$$

The charge in the ionization chamber is given by

$$\bar{D}_{air} = \frac{Q}{m}\left(\frac{\bar{W}}{e}\right)_{air} \qquad (4.3.68)$$

where
Q=46.3 nC (measured charge)

$(\bar{W}/e)_{air}$=33.97 J C^{-1} (mean energy expended to produce a charge of 1 C in air)
T=295 K, p=100.2 kPa (temperature and pressure at measurement point)
m=$v\rho$ (mass of air in the chamber volume v)

The density is given by

$$\rho = \rho_0 \frac{T_0 p}{T p_0} = 1.293 \frac{273.2 \cdot 100.2}{295 \cdot 101.3} \text{ kg m}^{-3} \tag{4.3.69}$$

The relation between absorbed dose in water and absorbed dose in air is, assuming no perturbation effects, given by

$$\bar{D}_{air} = D_{water} (\bar{s}_{water,air})^{-1} \tag{4.3.70}$$

The stopping power ratio $\bar{s}_{water,air}$ at the reference depth is 1.080 (Table 20, (IAEA, 2000)).

D_{water}=2.0 Gy (200 m.u.)

The air volume of the chamber is then

$$v = \frac{46.3 \cdot 10^{-9} \cdot 33.97 \cdot 1.080 \cdot 295 \cdot 101.3}{2.00 \cdot 1.293 \cdot 273.2 \cdot 100.2} = 7.17 \cdot 10^{-7} \text{ m}^3$$

Answer: The effective air volume is 0.72 cm^3.

4.3.5 Calorimetric and Chemical Dosimeters

Solution exercise 4.24.
The absorbed dose rate to the calorimeter is given by the relationship

$$\dot{D}_{water} = \Delta T c \tag{4.3.71}$$

where
ΔT =7.5·10^{-4} °C (increase in temperature)
c=4.186 kJ kg^{-1} °C (specific heat for water)
t=300 s (irradiation time)

The dose rate to water is given by

$$\dot{D}_{water} = I_Q N_{D,w} k_Q \tag{4.3.72}$$

where
$I_Q = I_m k_{Tp}$ (measured current corrected for pressure and temperature)

$N_{D,w}=1.758 \cdot 10^8$ Gy C^{-1} (calibration factor for absorbed to water at reference quality)
$k_Q=0.986$ (specific factor for chamber NE2581 to correct for the difference between the reference quality and the specific quality used, Q. Table 14 (IAEA, 2000))

Correction factor k_{Tp} for temperature and pressure is given by

$$k_{Tp} = \frac{p_0 T}{p T_0} \qquad (4.3.73)$$

where
$T_0=273.2$ K, $p_0=101.3$ kPa (temperature and pressure at calibration)
$T=295$ K, $p=100.2$ kPa (temperature and pressure at measurement)

The measured current, I_m, is then given by

$$I_m = \frac{\Delta T c}{t N_{D,w} k_Q k_{Tp}} \qquad (4.3.74)$$

Data inserted in Eq. (4.3.74) gives

$$I_m = \frac{7.5 \cdot 10^{-3} \cdot 4.186 \cdot 10^3}{300 \cdot 1.758 \cdot 10^8 \cdot 0.986} \frac{100.2 \cdot 273.2}{101.3 \cdot 295} = 0.553 \text{ nA}$$

Answer: The ionization current is 0.55 nA.

Solution exercise 4.25.
The absorbed dose to the Fricke dosimeter is obtained by the relation

$$\bar{D}_{Fe3+} = \frac{\Delta A}{\rho l \epsilon G_{Fe3+}} \qquad (4.3.75)$$

Absorbed dose to water is assuming charged particle equilibrium given by

$$D_{water} = \bar{D}_{Fe3+} \frac{(\mu_{en}/\rho)_{water}}{(\mu_{en}/\rho)_{FeSO_4}} \qquad (4.3.76)$$

where
$\Delta A=0.350$ ODU (absorbance of Fe^{3+} in the solution)
$\rho = 1.024 \cdot 10^3$ kg m^{-3} (density of the solution)
$l=10$ mm (light path length)
$\epsilon=220$ m^2 mol^{-1} (difference in molar linear absorption coefficient for ferric and ferrous ions)
$G_{Fe3+}=1.62 \cdot 10^{-6}$ molkg^{-1} Gy^{-1} (radiation chemical yield of ferric ions)
$(\mu_{en}/\rho)_{water} = 0.00296$ m^2 kg^{-1} (mass energy absorption coefficient for water)
$(\mu_{en}/\rho)_{FeSO_4} = 0.00295$ m^2 kg^{-1} (mass energy absorption coefficient for ferrous

sulfate solution)

Combining Eq.(4.3.76) and (4.3.75), inserting data, and dividing by the irradiation time of 50 h, gives the dose rate.

$$\dot{D}_{\text{water}} = \frac{0.350 \cdot 0.00296}{1.024 \cdot 10^3 \cdot 0.01 \cdot 220 \cdot 1.62 \cdot 10^{-6} \cdot 0.00295 \cdot 50} = 1.92\,\text{Gy}\,\text{h}^{-1}$$

Answer: The absorbed dose rate is $1.92\,\text{Gy}\,\text{h}^{-1}$.

Solution exercise 4.26

The absorbed dose to the ferrous sulfate dosimeter is obtained by the relation

$$\bar{D}_{(Fe3+)} = \frac{\Delta A}{\rho_{\text{sol}}\, l\epsilon G_{\text{Fe3+}}} \tag{4.3.77}$$

Absorbed dose to water is given by

$$D_{\text{water}} = \bar{D}_{\text{Fe3+}} S_{\text{water,FeSO}_4} \tag{4.3.78}$$

where
ΔA=0.832 ODU (absorbance of Fe^{3+} in the solution)
ρ_{sol} = $1.024 \cdot 10^3$ kg m^{-3} (density of the solution)
l=10 mm (light path length)
$\epsilon G_{\text{Fe3+}}$=352$\cdot 10^{-6}$ m^2 kg^{-1} Gy^{-1}

The energy of the electrons at the reference depth is obtained by using the relation (IAEA, 2000)
$$E_{\text{ref}} = 0.07 + 1.027 R_{50} - 0.0048(R_{50})^2 \tag{4.3.79}$$
Inserting R_{50}=20 g cm^{-2} gives

$$E_{\text{ref}} = 0.07 + 1.027 \cdot 20 - 0.0048(20)^2 = 18.7\,\text{MeV}$$

This gives
$(S_{\text{el}}/\rho)_{\text{water}}$=0.2038 MeV m^2 kg$^-$1 (mass collision stopping power for water)
$(S_{\text{el}}/\rho)_{\text{FeSO}_4}$=0.2029 MeV m^2 kg$^-$1 (mass collision stopping power for ferrous sulfate solution).

Combining Eq. (4.3.77) and (4.3.78), inserting data, and dividing by the irradiation time of 100 min gives the absorbed dose rate to water.

$$\dot{D}_{\text{Fe3+}} = \frac{0.832 \cdot 0.2038}{1.024 \cdot 10^3 \cdot 0.01 \cdot 352 \cdot 10^{-6} \cdot 100 \cdot 0.2029} = 2.318\,\text{Gy}\,\text{min}^{-1}$$

The ionization chamber measured at the same point, obtained a charge of 3.34 nC (T=293 K, p=100.2 kPa) after an irradiation of 60 s. Correcting for temperature and pressure, the calibration factor at (T=295 K, p=101.3 kPa) is given by

$$k = \frac{2.318 \cdot 100.2 \cdot 295}{3.34 \cdot 101.3 \cdot 293} = 0.691 \text{ Gy in water per nC}$$

Answer: The calibration factor is 0.69 Gy in water per nC.

Bibliography

Almond P. R., Biggs P. J., Coursey B. M., et al. (1999). AAPM's TG-51 protocol for clinical reference dosimetry of high-energy photon and electron beams. Medical Physics, 26, 1847-1870.

Brahme A. (1975). Simple relations for the penetration of high energy electron beams in matter. SSI:1975-011. Internal report. Stockholm, Sweden. National Institute of Radiation protection.

Burlin T. E. (1966). A general theory of cavity ionization. Brit J Radiol, 39, 727-734.

International Commission on Radiological Units and Measurements. (1984). Radiation Dosimetry: Electron Beams with Energies Between 1 and 50 MeV. ICRU Report 35. Bethesda:ICRU Publications.

Nahum A. E. (1978). Water/air stopping power ratios for megavoltage photon and electron beams. Physics in Medicine and Biology, 23, 24-38.

5 Radiation Biology

5.1 Definitions and Relations

The material below is mainly related to cell survival and clinical therapeutic applications are not included. The cell survival models are introduced without any discussion of the biology behind them.

Cell survival

Cell survival can have different meanings depending on the context. *Differentiated cells* that do not divide can be regarded as dead when they lose their specific function. *Proliferating cells* that are dividing cells can be regarded as dead when they lose their reproductive capacity. They may divide a couple of times, but they have lost their *clonogenic* capacity.

Cells that die when trying to divide are undergoing a so called *mitotic death*. Another type of death is *apoptosis*, also called programmed cell death. This type of death leads to changes in the function of the cell as nuclear fragmentation, chromatin condensation and chromosomal DNA fragmentation. Both types of cell death will lead to the loss of cell reproduction.

Plating efficiency

A common way to measure cell survival is to grow single cells in a nutritional liquid on dishes. These cells will then grow into large colonies that can be seen by the naked eye if the cells still have their reproductive capacity. By calculating the number of colonies for cells irradiated with different absorbed doses and under different conditions, it is possible to obtain cell survival curves. It is then important to have a reference dish with unirradiated cells. *Plating efficiency* is defined as the percentage of cells that grow into colonies.

$$PE = \frac{\text{Number of colonies counted}}{\text{Number of cells seeded}} \qquad (5.1.1)$$

Surviving fraction is then defined as

$$SF = \frac{\text{Number of colonies counted}}{\text{Number of cells seeded} \cdot PE} \qquad (5.1.2)$$

Cell survival models

Exponential dose response model. Cell survival data are often plotted as the logarithm of the surviving fraction. This means that a linear curve indicates an exponential relation between absorbed dose and survival.

$$SF = \frac{N}{N_0} = e^{-D/D_0} \qquad (5.1.3)$$

where SF is the ratio of the number of surviving cells N to the initial number of cells N_0. D is the absorbed dose and D_0 is the absorbed dose that reduces the number of cells to 37% of the initial number. This is sometimes written as D_{37}. $1/D_0$ gives the slope of the curve. In Fig. 5.1 an exponential survival curve is included with $D_0 = 0.96$ Gy.

An absorbed dose of D_0 corresponds to on average one hit per cell. That not all cells are killed is due to statistics. Some cells will get more than one hit and some will not be hit. The distribution of hits is given by the binomial distribution, which for a large number of cells may be approximated by the Poisson distribution. The the probability to obtain n hits in a cell is then given by

$$P(n) = \frac{e^{-x} x^n}{n!} \tag{5.1.4}$$

where x is the average number of hits and n is the specific number of hits. If each hit results in a cell inactivation, then the probability of survival is the probability of not being hit, $P(0)$. Thus with $x=1$ and $n=0$

$$P(0) = \frac{e^{-1} 1^0}{0!} = e^{-1} = 0.37 \tag{5.1.5}$$

D_0 is also often called the *mean lethal dose*.

Exponential cell survival is more common for densely ionizing radiation like neutrons and ions. Sometimes this curve is interpreted such that there is enough energy deposited in a hit to inactivate the cell (single-hit, single-target), and that there is no repair of the cells.

Survival curves with shoulder. Experimental survival curves often have a shoulder followed by an exponential decrease. Several mathematical models have been derived to fit the experimental data and there have been different suggestions on how to interpret these results from a more fundamental point of view. This is under debate and will not be included here. However, there is probably some relation between the response and the energy deposited by each track within some or several important targets (DNA, cell nucleus,...). The possibility to repair the initial damages is also of importance for the shape of the survival curves.

Single-hit multi-target model. The first interpretation of the shoulder was by using the "single-hit multi-target" model. In this model it is assumed that to inactivate the cell there is a need for one hit in n targets. The survival fraction will then be given by

$$SF = 1 - (1 - e^{-D/D_0})^n \tag{5.1.6}$$

where D_0 is the dose to reduce cell survival to 37% along the log-linear slope of the curve. n is the extrapolation number, that is the value of the y-axis when $D=0$. The interpretation of hits and targets is probably not correct, but the equation may describe the shape of the survival, in particular at high doses.

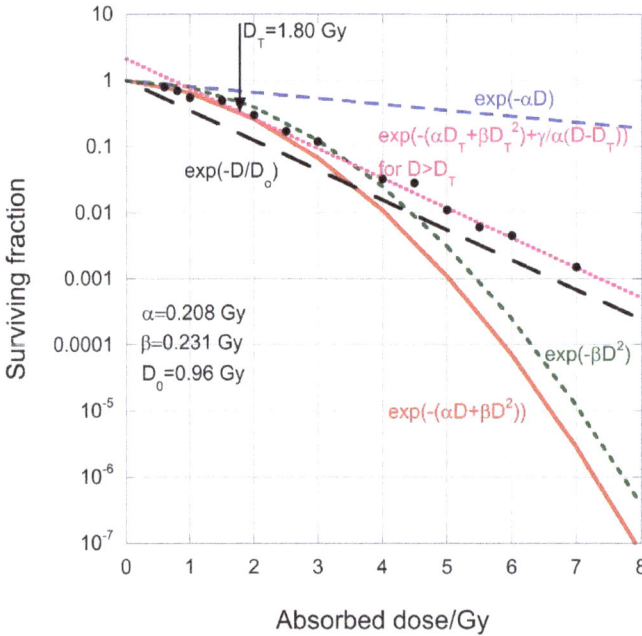

Figure 5.1: Different cell survival curves compared with experimental data for mammalian cells (Puck and Markus, 1956). The LQ-model shows a good agreement for low doses below 2-3 Gy. With increasing doses the difference between experiment and the LQ-model increases. In the figure the separate components $e^{-\alpha D}$ and $e^{-\beta D^2}$ are also included. The absorbed dose where these curves intersect gives the α/β-ratio (0.90 Gy). For large absorbed doses the experimental data can be fitted to a line in the lin-log diagram. The slope of this line is given by $1/D_0$, where D_0 (0.96 Gy) is the mean lethal dose for high absorbed doses. The LQ-L-model gives a good agreement with the experiments for the full dose range. For doses below D_T (1.80 Gy) the LQ-L-model agrees with the LQ-model.

Linear quadratic (LQ) model. The most common model used today in radiotherapy is the linear quadratic model.

$$SF = e^{-(\alpha D + \beta D^2)} \tag{5.1.7}$$

where α and β are fitting parameters that depend on the type of cells and radiation quality. α gives the slope of the initial part of the curve (See Fig. 5.1). The relation has sometimes been interpreted such that an inactivation can be produced by either a single track ($P = \alpha D$), or by two independent tracks ($P = \beta D^2$). However, this explanation does not include any discussion of repair and is not accepted today. This equation is the basis for many of the relations used to consider fractionation regimes in radiotherapy. However, one drawback of this curve is that the curvature increases continuously with absorbed dose. This not obtained experimentally and it is also difficult to find support from fundamental radiobiology. The difference between the model and experimental survival curves is rather small at absorbed doses around and below 2 Gy, which has been a normally used daily dose. See Fig. 5.1. However, recently it has become more common to use higher doses per fraction in radiotherapy and as such the difference between model and experiment is not acceptable. Lately there have thus been attempts to improve this model.

Linear quadratic linear (LQ-L) model. This model fits the first part of the LQ-model with a linear part at high doses (Astrahan, 2008). The model gives the transition between the two parts as

$$SF = e^{-(\alpha D + \beta D^2)} \text{ for } D < D_T \tag{5.1.8}$$

and

$$SF = e^{-(\alpha D_T + \beta D_T^2 + \gamma(D - D_T))} \text{ for } D \geq D_T \tag{5.1.9}$$

where γ is the cell kill per Gy in the final linear part of the curve at large absorbed doses.

The γ/α ratio can be calculated from the slope of the tangent at D_T and the α/β term

$$\gamma/\alpha = 1 + (2D_T/(\alpha/\beta)) \tag{5.1.10}$$

The value of D_T may be obtained from experiments. This model can better describe cell survival curves over a large dose range (see Fig. 5.1).

Repairable conditionally repairable (RCR) model. The RCR-model (Lind et al, 2003) tries to include the repair of the cells into the model. The survival fraction in this model is given by

$$SF = e^{-aD} + bDe^{-cD} \tag{5.1.11}$$

where a, b and c are the parameters of the model. e^{-aD} corresponds to the undamaged cells. bDe^{-cD} corresponds to the fraction of cells that have been damaged and subsequently repaired. The model separates between two types of damage. One is the potentially repairable damage, that may also be lethal, and thus non-repaired or mis-repaired. The other type is the conditionally repairable damage that may lead to an apoptotic response if not repaired. This expression for survival gives a good fit to experimental data over the full range of doses, from the very low, where there may be a hypersensitivity, to the exponential part at very high doses.

Fig. 5.2 shows a comparison between the LQ-model and the RCR model fitted to experimental cell survival data. The figure to the left shows that the RCR-model gives a lower survival at low doses than the LQ-model and thus provides a better means of simulating hypersensitivity. The figure to the right makes a comparison over a larger dose range and shows that the RCR-model gives a straighter line in the lin-log diagram than the LQ-model, which is in good agreement with experimental data. The

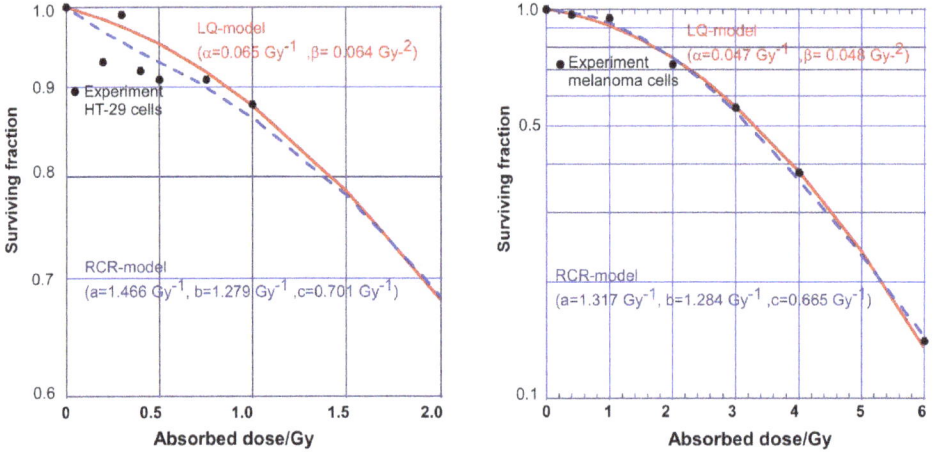

Figure 5.2: Survival fraction after irradiation with ^{60}Co for HT-29 cells (left figure) and human melanoma cells (right figure). The experimental data are fitted to the LQ model and the RCR model.

data in the figure are taken from Persson (Persson, 2002).

Oxygen enhancement ratio, OER.
Survival curves for cells exposed to radiation in the presence and absence of oxygen may be different. The ratio of doses under hypoxic and aerated situations to achieve the same biological effect is called oxygen enhancement ratio, OER.

$$OER = \frac{D_{\text{hypoxic}}}{D_{\text{aerated}}} \tag{5.1.12}$$

OER is dependent on radiation quality and is typical 2.5 to 3.5 for photons, 1.5 for neutrons and 1.0 for α-particles. Hypoxia is an important factor to consider in radiotherapy.

Relative biological effectiveness, RBE.
Different radiation qualities may have different effectiveness. The *relative biological effectiveness*, RBE, of some test radiation compared with a reference radiation is defined as

$$RBE = \frac{D_{\text{ref}}}{D_{\text{test}}} \tag{5.1.13}$$

where D_{ref} is the absorbed dose at the reference radiation quality at a certain biological endpoint and D_{test} is the absorbed dose at the test radiation quality at the same biological endpoint.

It is important to consider that RBE is defined for a certain biological endpoint, and may differ significantly when determining RBE at a cell survival for e.g. 50% or 10%. The radiation quality is often expressed in terms of LET (Linear Energy Transfer).

5.2 Exercises in Radiation Biology

Exercise 5.1. 100 unirradiated cells were seeded on a dish to determine the cell survival. After incubating for two weeks 75 colonies were obtained. On another dish 100 cells were seeded and irradiated with an absorbed dose of 2.0 Gy. After two weeks 20 colonies were counted. Calculate the D_0 value, assuming an exponential survival curve.

Exercise 5.2. Bacteria are irradiated by placing them in a radioactive solution with a β-emitting radionuclide. The number of surviving bacteria decreases with time (absorbed dose) approximately exponentially. This is assumed to confirm that the one-hit theory holds. 50% of the cells survive at an absorbed dose of 800 Gy. Estimate the sensitive volume in which the target theory assumes that one ionization kills the bacterium. The density of the bacterium $= 1.35 \cdot 10^3$ kg m^{-3}, $W_0 = 110$ eV. W_0 is the energy needed to inactivate the bacterium.

Exercise 5.3. A cell population is irradiated with 40.0 Gy (single irradiation). D_{10} (i e the absorbed dose that reduces the population to 10%) is 4.0 Gy, when the cells are irradiated at normal oxygen pressure.
a) Calculate the relative survival if it is assumed that the survival can be assumed to follow a single-hit, single target model without any repair.
b) Calculate the survival if 1% of the population is hypoxic (i.e. with low concentration of oxygen). D_{10} for hypoxic cells is supposed to be twice as large as for corresponding cells with normal oxygen pressure.

Exercise 5.4. A tumor has $1.0 \cdot 10^8$ cells. The tumor is irradiated with neutrons which means that one may assume that the survival curve is exponential. The D_0-value is 1.9 Gy. What absorbed dose is needed to obtain a probability of 1% that no cell survives? Assume that Poisson statistics may be used.

Exercise 5.5 The cell survival is often described by the linear quadratic model. A tumor with an α/β-ratio=10 Gy and β=3.2·10^{-2} Gy^{-2} is going to be treated. Normal cells will also be irradiated. These cells have an α/β-ratio = 3.0 Gy and β=5.1·10^{-2} Gy^{-2}. The treatment schedule implies 2.0 Gy every day for 30 days both for tumor cells and for normal cells. Calculate the survival of the two cell types. If the absorbed dose per fraction instead is 1.0 Gy, how many fractions are then necessary in order to obtain the same survival of the tumor cells? Which is now the survival of the normal cells? If the absorbed dose instead is given in one treatment, which absorbed dose is needed to give the same tumor cell survival? Which is now the survival of the normal cells? Assume full repair between the treatments.

Exercise 5.6. Cells are irradiated with neutrons and in this case their survival can be described using a single hit, single target model with D_0=1.7 Gy. When irradiating with photons from a ^{60}Co-source, the survival curve can be described using the linear-quadratic-model with α/β=3.0 Gy and α=0.15 Gy^{-1}. Calculate RBE for a survival of 50% respectively 10% assuming that the reference quality is ^{60}Co photons.

Exercise 5.7. Cells are irradiated with neutrons and in this case their survival can be described using an exponential survival curve with D_0=1.5 Gy. When irradiating with photons the the survival curve has a shoulder and then an exponential part with a slope equal to the slope for the neutron irradiation. This curve is typical for what sometimes was described as a "single hit, multiple target model". When extrapolating the linear part of the curve to the absorbed dose, D=0, the extrapolation number n=3. Calculate the cell survival for a photon dose of 2.0 Gy. Which neutron dose would give the same cell survival? Assume that the patient is treated with photons in 30 fractions with 2.0 Gy per fraction. Calculate the cell survival in this situation assuming full repair between the fractions. If this survival should be obtained in a single dose, which absorbed dose is needed, for neutrons and photons?

Exercise 5.8. Chinese hamster cells irradiated with x rays show a cell survival according to Fig. 5.3. The data may be fitted with either the LQ model or the LQ-L model. Compare these two models with the experimental data, and plot them in a figure together with the experimental data. α=0.17 Gy^{-1}, β=0.06 Gy^{-2} and D_T=2α/β Gy. The cells are treated with a total absorbed dose of 20 Gy in fractions of a) 2 Gy and b) 10 Gy. Calculate the survival using the two models.

Exercise 5.9. Human melanoma cells are irradiated with ^{60}Co-γ-rays and ^{10}B ions with an LET of 160 eV nm^{-1}. The experimental survival curves are fitted with the LQ-model and the RCR-model. The parameters for the models are tabulated in Table 6.1.

Calculate RBE using the two models for a cell survival of 0.8 and 0.1.

Table 5.1: Parameters for the LQ- and the RCR models (Persson, 2002).

Radiation quality	LQ model	RCR model
^{60}Co	$\alpha = 0.047, \beta = 0.048$	a=1.317, b=1.284, c=0.665
^{10}B	$\alpha = 0.712, \beta = 0.226$	a=1.860, b=1.508, c=1.856

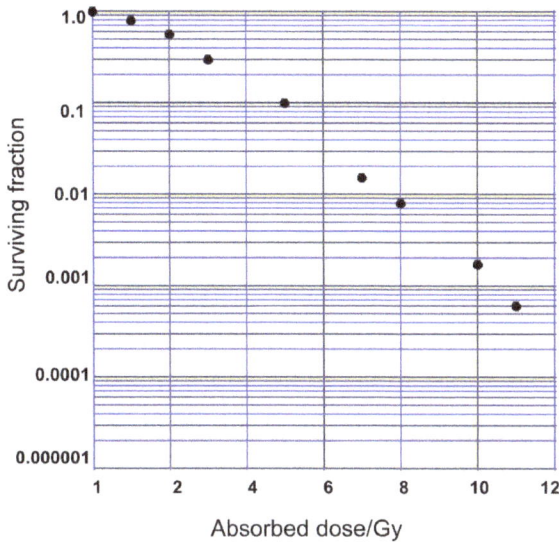

Figure 5.3: Cell survival for Chinese Hamster Cells.

5.3 Solutions in Radiation Biology

Solution exercise 5.1.

The plating efficiency (PE) is defined as

$$PE = \frac{\text{Number of colonies counted}}{\text{Number of cells seeded}} \tag{5.3.1}$$

The surviving fraction (SF) is obtained by

$$SF = \frac{\text{Number of colonies counted}}{\text{Number of cells seeded} \cdot PE} \tag{5.3.2}$$

The value of D_0 is obtained from

$$SF = e^{-D/D_0} \tag{5.3.3}$$

where SF is the surviving fraction, D is the absorbed dose and D_0 is the dose that reduces the survival to 37% of the initial number of cells.

Data:

Unirradiated cells: Number of cells seeded=100, number of colonies=75

Irradiated cells: Number of cells seeded=100, number of colonies=20

Data inserted gives

$$PE = \frac{75}{100} = 0.75$$

Surviving fraction

$$SF = \frac{20}{100 \cdot 0.75} = 0.267$$

This inserted in Eq. (5.3.3) together with D=2.0 Gy gives

$$0.267 = e^{-2.0/D_0}$$

This gives D_0 = 1.51 Gy.

Answer: The mean lethal dose, D_0, is 1.5 Gy.

Solution exercise 5.2.

The energy E that is released in a volume v m^3 is given by

$$E = D\rho v \tag{5.3.4}$$

Number of "hits", n, in volume v is given by

$$n = \frac{D\rho v}{W_0} \tag{5.3.5}$$

where D is the absorbed dose and ρ the density of bacteria. W_0 is the energy needed to inactivate the bacterium.

For an absorbed dose D_0 (D_{37}) there is on average one hit (n=1) per bacterium if the survival curve is exponential. This gives

$$v = \frac{W_0}{D_0 \rho} \tag{5.3.6}$$

For these bacteria D_{50} is 800 Gy. The survival fraction for an exponential survival is given by

$$SF = e^{-D/D_0} \tag{5.3.7}$$

Data inserted gives

$$0.5 = e^{-800/D_0}$$

and

D_0 = 1154 Gy

With ρ=1.35·10^3 kg m^{-3} and W_0=110 eV inserted in Eq. (5.3.6), v is given by

$$v = \frac{110 \cdot 1.602 \cdot 10^{-19}}{1154 \cdot 1.35 \cdot 10^3} = 1.13 \cdot 10^{-23} \text{ m}^3$$

Answer: The active volume of the bacteria is $1.13 \cdot 10^{-23}$ m^3.

Solution exercise 5.3.

The survival fraction of cells is given by

$$SF = e^{-D/D_0} \tag{5.3.8}$$

With D_{10}=4.00 Gy for oxygenated cells, then D_0 is obtained from

$$0.1 = e^{-4.0/D_0}$$

Solving the equation gives $D_0 = -4.0/\ln 0.1$. Inserting this in Eq. (5.3.8) and setting D=40 Gy gives

$$SF = e^{10 \ln 0.1} = 10^{-10}$$

If the cells are hypoxic then D_{10}=8.00 Gy. The relative survival is then given by

$$SF = e^{5 \ln 0.1} = 10^{-5}$$

If in a population of cells, 1 % are hypoxic, then the total survival is given by

$$SF = 0.99 \cdot 10^{-10} + 0.01 \cdot 10^{-5} = 10^{-7}$$

The survival is dominated by the hypoxic cells.

Answer: The survival fraction is 10^{-10} for oxygenated cells and 10^{-7} for a population consisting of 99% oxygenated cells and 1% hypoxic cells.

Solution exercise 5.4.

The survival fraction of cells is given by

$$SF = N/N_0 = e^{-D/D_0} \tag{5.3.9}$$

where N is the average number of surviving cells after an absorbed dose of D Gy and N_0 is the number of cells before irradiation. Assume that the cell survival follows a Poisson distribution. Then the probability that n cells survive if there are N_0 cells initially and on average N cells survive is given by

$$P(n) = \frac{e^{-N} N^n}{n!} = \frac{e^{-N_0 e^{-D/D_0}} (N_0 e^{-D/D_0})^n}{n!} \tag{5.3.10}$$

The probability that no cell will survive is obtained by setting n=0. Thus

$$P(0) = \frac{e^{-N_0 e^{-D/D_0}} (N_0 e^{-D/D_0})^0}{0!} = e^{-N_0 e^{-D/D_0}} \tag{5.3.11}$$

Data:
D_0=1.9 Gy (mean lethal dose)
$P(0)$=0.01 (probability of no cell survival)
$N_0 = 10^8$ (number of initial cells)

Data inserted in Eq. (5.3.11) gives

$$0.01 = e^{-10^8 e^{-D/1.9}}$$

Taking the logarithm of the equation gives

$$\ln 0.01 = -10^8 e^{-D/1.9}$$

and D=32.1 Gy.

Answer: The absorbed dose needed to get a survival probability of 1% is 32 Gy.

Solution exercise 5.5.
Cell survival after one treatment for an absorbed dose D with the LQ-model is given by

$$SF = e^{-(\alpha D + \beta D^2)} \qquad (5.3.12)$$

After N treatments the survival is

$$SF_N = [e^{-(\alpha D + \beta D^2)}]^N \qquad (5.3.13)$$

Data:
D=2.0 Gy (absorbed dose)
N=30 (number of treatments)

Tumor cells
$\beta = 3.2 \cdot 10^{-2}$ Gy^{-2}, α/β=10 Gy $\Rightarrow \alpha = 0.32$ Gy^{-1}

Normal cells
$\beta = 5.2 \cdot 10^{-2}$ Gy^{-2}, α/β=3.0 Gy $\Rightarrow \alpha = 0.156$ Gy^{-1}

Data inserted in Eq. (5.3.13) gives

Tumor cells:

$$SF_{30} = [e^{(-0.32 \cdot 2.0 - 3.2 \cdot 10^{-2} 2^2)}]^{30} = 9.860 \cdot 10^{-11}$$

Normal cells:

$$SF_{30} = [e^{(-0.156 \cdot 2.0 - 5.2 \cdot 10^{-2} 2^2)}]^{30} = 1.679 \cdot 10^{-7}$$

The ratio of the survival tumor cells to normal cells is $9.860 \cdot 10^{-11}/1.679 \cdot 10^{-7} = 5.87 \cdot 10^{-4}$.

If instead the absorbed dose per fraction is 1.0 Gy, the number of treatments for the same survival for tumor cells is given by

$$9.860 \cdot 10^{-11} = [e^{-(0.32 \cdot 1.0 + 3.2 \cdot 10^{-2} 1.0^2)}]^N = [e^{-0.352}]^N$$

and

$$N = \ln 9.860 \cdot 10^{-11}/(-0.352) = 65.45$$

Thus 65.45 treatments and 65.45 Gy are needed. This can compared with the absorbed dose 30x2 Gy = 60 Gy that was needed with an absorbed dose of 2.0 Gy per treatment.

The survival fraction of normal cells will in this situation be

$$SF_{65} = [e^{-(0.156 \cdot 1.0 + 5.2 \cdot 10^{-2} 1.0^2)}]^{65.45} = 1.224 \cdot 10^{-6}$$

The ratio of the survival tumor cells to normal cells is in this case $9.860 \cdot 10^{-11}/1.224 \cdot 10^{-6} = 8.06 \cdot 10^{-5}$.

This gives a better ratio than for a treatment with 2.0 Gy per treatment as more normal cells will survive. However repopulation is not included in the calculations.

If the total absorbed dose is given in a single treatment, then with the same survival of the tumor cells, the absorbed dose will be given by

$$9.860 \cdot 10^{-11} = [-e^{-(0.32 \cdot D + 3.2 \cdot 10^{-2} D^2)}]$$

and

$$-\ln(9.860 \cdot 10^{-11}) = 0.32 \cdot D + 3.2 \cdot 10^{-2} D^2$$

giving

$$D = -\frac{0.32}{0.032 \cdot 2} \pm \sqrt{(\frac{0.32}{0.032 \cdot 2})^2 - \frac{\ln(9.860 \cdot 10^{-11})}{0.032}} = 22.3 \text{ Gy}$$

The survival fraction of normal cells will then be

$$SF = [e^{-(0.156 \cdot 22.3 + 5.2 \cdot 10^{-2} 22.3^2)}] = 1.81 \cdot 10^{-13}$$

The ratio of the survival tumor cells to normal cells is $9.860 \cdot 10^{-11}/1.81 \cdot 10^{-13} = 543$. Thus in this case the survival of the normal cells is less than for the tumor cells.

Answer: The ratio between the survival of tumor cells and normal cells is $5.9 \cdot 10^{-4}$ with 2.0 Gy per treatment. With 1.0 Gy per treatment a total absorbed dose of 65.4 Gy is needed and the ratio between the survival of tumor cells and normal cells is $8.1 \cdot 10^{-5}$. For a single treatment an absorbed dose of 22.3 Gy is needed and the ratio between the survival of tumor cells and the normal cells is 543.

Solution exercise 5.6.

Neutrons:

The cell survival when irradiating with neutrons will give an exponential curve as the survival follows the single hit-single target model. Thus

$$SF = e^{-D/D_0} \qquad (5.3.14)$$

With D_0=1.7 Gy the doses for 10% and 50% survival will be

$$0.1 = e^{-D_{10}/1.7}$$

and

$$0.5 = e^{-D_{50}/1.7}$$

respectively. This gives D_{10}=3.91 Gy and D_{50}=1.18 Gy.

Photons:

The survival follows the linear quadratic model. Thus the survival will be given by

$$SF = e^{-(\alpha D + \beta D^2)} \qquad (5.3.15)$$

where α=0.15 Gy^{-1}, α/β=3.0 Gy and β=0.05 Gy^{-2}. This inserted in Eq. (5.3.15) gives the doses for 10 % and 50 % survival respectively.

$$0.1 = e^{-(0.15 D_{10} + 0.05 D_{10}^2)}$$

and

$$0.5 = e^{-(0.15 D_{50} + 0.05 D_{50}^2)}$$

Solving the equations give D_{10}=5.45 Gy and D_{50}=2.51 Gy

RBE is defined as the ratio of doses for the same effect or cell survival. Thus

$$RBE_{10} = 5.45/3.91 = 1.39$$

and

$$RBE_{50} = 2.51/1.18 = 2.13$$

Answer: RBE for 10% survival is 1.4 and for 50% survival 2.1.

Solution exercise 5.7.

Photons:

The cell survival is, assuming a single-hit multi-target model, obtained from the relation

$$SF = 1 - (1 - e^{-D/D_0})^n \qquad (5.3.16)$$

where D_0=1.5 Gy (dose to reduce cell survival to 37% along the linear slope of the curve) and n=3 (extrapolation number).

Eq. (5.3.16) gives the survival fraction after an absorbed dose of 2.0 Gy

$$SF = 1 - (1 - e^{-2.0/1.5})^3 = 0.6006$$

With 30 fractions the survival fraction will be

$$SF = (1 - (1 - e^{-2.0/1.5})^3)^{30} = 2.284 \cdot 10^{-7}$$

If the absorbed dose is to be given in a single fraction, the absorbed dose for the same survival as from 30 fractions, will be obtained from

$$2.284 \cdot 10^{-7} = 1 - (1 - e^{-D/1.5})^3$$

D=24.6 Gy

This can be compared to 60 Gy needed in a 30 fraction treatment of 2 Gy.

Neutrons

The survival for neutrons is assumed to be exponential. Then the absorbed dose for a survival of 0.601 in one fraction is given by

$$0.601 = e^{-D/1.5}$$

D=0.765 Gy.

To obtain a survival fraction of $2.284 \cdot 10^{-7}$ for neutrons in a single fraction the absorbed dose needed is given by

$$2.284 \cdot 10^{-7} = e^{-D/1.5}$$

D=22.9 Gy.

Answer: The cell survival fraction for an absorbed dose of 2.0 Gy is 0.60 with photons. To obtain the same survival with neutrons an absorbed dose of 0.77 Gy is needed. The survival after 30 equal fractions with photons is $2.28 \cdot 10^{-7}$. To obtain the same survival in a single fraction, an absorbed dose of 25 Gy is needed for photons and 23 Gy for neutrons.

Solution exercise 5.8.

Assume that the survival follows the LQ model. The survival fraction will be given by

$$SF_{LQ} = e^{-(\alpha D + \beta D^2)} \tag{5.3.17}$$

where α=0.170 Gy^{-1} and β=0.06 Gy^{-2}.

The LQ-L model gives the survival fraction according to LQ model for doses below a dose D_T and for higher doses according to

$$SF_{LQ-L} = e^{-(\alpha D_T + \beta D_T^2 + \gamma(D - D_T))} \tag{5.3.18}$$

where

$$\gamma = \alpha[1 + (2D_T/(\alpha/\beta))] \tag{5.3.19}$$

In this case, and it holds for many survival curves, $D_T = 2\alpha/\beta$. Thus

$$\gamma = 0.17[1 + (2 \cdot 2)] = 0.85$$

Data inserted in Eqs. (5.3.17) and (5.3.18) gives the survival curves. The result is shown in Fig. 5.4. It is clear that for absorbed doses below around 5 Gy, both models agree with the experiments but for large absorbed doses the LQ-model underestimates the survival.

For a dose per fraction of 2.0 Gy, both models give the same result, and calculations can be made with the LQ model. An absorbed dose of 20 Gy in 2.0 Gy fractions will give a survival of

$$SF_{10} = (e^{-(0.170 \cdot 2 + 0.06 \cdot 2.0^2)})^{10} = 3.03 \cdot 10^{-3}$$

With a dose per fraction of 10 Gy, the two models will give different results.

LQ model

The survival will for a total dose of 20 Gy be

$$SF_{LQ,2} = (e^{-(0.170 \cdot 10 + 0.06 \cdot 10^2)})^2 = 2.05 \cdot 10^{-7}$$

LQ-L model

The corresponding survival will be with D_T=2·0.170/0.06=5.67 Gy.

$$SF_{LQ-L,2} = (e^{-(0.17 \cdot 5.67 + 0.06 \cdot 5.67^2 + 0.85(10 - 5.67))})^2 = 1.95 \cdot 10^{-6}$$

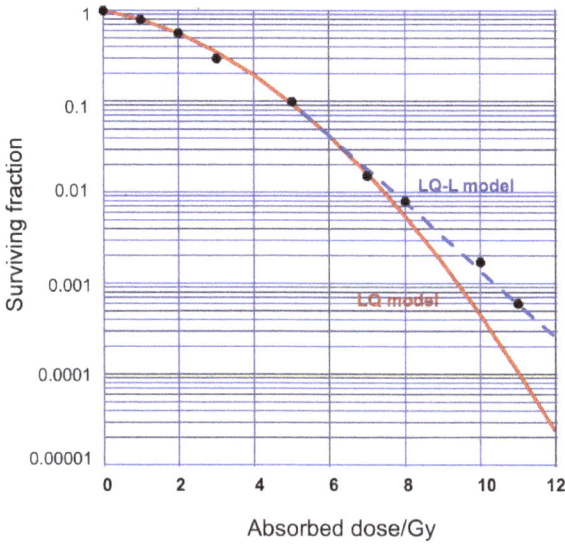

Figure 5.4: Cell survival for Chinese Hamster Cells. Experimental data and applied LQ model and LQ-L model respectively.

When treating patients with few fractions of large doses the LQ-model will give large differences compared to what is expected from cell survival data.

Answer: The survival with 10 fractions of 2.0 Gy is $3.03 \cdot 10^{-3}$ for both models. With two fractions of 10 Gy the survival is $2.05 \cdot 10^{-7}$ with the LQ- model and $1.95 \cdot 10^{-6}$ with the LQ-L model.

Solution exercise 5.9.
^{60}**Co**

LQ model

$$SF_{LQ} = e^{-(\alpha D + \beta D^2)} \tag{5.3.20}$$

With α =0.047 Gy^{-1} and β =0.048 Gy^{-2}, the absorbed dose to achieve a survival rate of 80% is obtained by

$$0.8 = e^{-(0.047D + 0.048D^2)}$$

Solving the equation gives $D_{0.8}$=1.72 Gy.

A cell survival rate of 10% gives correspondingly $D_{0.1}$=6.45 Gy.

<u>RCR model</u>

$$SF_{RCR} = e^{-aD} + bDe^{-cD} \qquad (5.3.21)$$

With a=1.317 Gy^{-1}, b=1.284 Gy^{-1} and c=0.665 Gy^{-1}, the absorbed dose to get a survival rate of 80% is obtained by

$$0.80 = e^{-1.317D} + 1.284De^{-0.665D}$$

Solving the equation numerically gives $D_{0.8}$=1.76 Gy.

A cell survival of 10% gives correspondingly $D_{0.1}$=6.70 Gy.

^{10}B ions

<u>LQ model</u>

With α =0.712 Gy^{-1} and β =0.226 Gy^{-2}, the absorbed dose to get a survival of 80% is obtained by

$$0.8 = e^{-(0.712D+0.226D^2)}$$

Solving the equation gives $D_{0.8}$=0.287 Gy.

A cell survival of 10 % gives correspondingly $D_{0.1}$=1.984 Gy.

<u>RCR model</u>

With a=1.860 Gy^{-1}, b=1.508 Gy^{-1} and c=1.856 Gy^{-1}, the absorbed dose to get a survival of 80% is obtained by

$$0.80 = e^{-1.860D} + 1.508De^{-1.856D}$$

Solving the equation numerically gives $D_{0.8}$=0.346 Gy.

A cell survival of 10% gives correspondingly $D_{0.1}$=1.986 Gy.

The RBE can now be determined

<u>LQ model</u>
$RBE_{0.8}$=1.72/0.287=6.0

$RBE_{0.1}$=6.45/1.984=3.3

RCR model
$RBE_{0.8}=1.76/0.346=5.1$

$RBE_{0.1}=6.70/1.986=3.4$

Answer: The RBE for 80% cell survival is 6.0 using the LQ-model and 5.1 using the RCR model. The RBE for 10% cell survival is 3.3 using the LQ-model and 3.4 using the RCR model.

Bibliography

Astrahan M. (2008). Some implications of linear quadratic-linear radiation dose-response with regard to hypofractionation. Medical Physics, 35, 4161-4172.

Edgren M. R., Lind B. K., Persson L. M., et al. (2003). Repairable-conditionally repairable damage model based on dual Poisson processes. Radiation Research, 160, 366-375.

Persson L. M. (2002). Cell survival at low and high ionization densities investigated with a new model. Thesis. Stockholm, Sweden. Stockholm university.

Puck T. T. and Markus P. L. (1956). Action of x-rays on mammalian cells. Journal of Experimental Medicine, 103, 653-666.

6 Radiation Protection and Health Physics

6.1 Definitions and Equations

6.1.1 Dose concepts

Equivalent dose

Equivalent dose is defined as

$$H_{T,R} = w_R D_{T,R} \tag{6.1.1}$$

where w_R is the radiation weighting factor and $D_{T,R}$ is the mean absorbed dose from radiation R in tissue or organ T. Numerical values of w_R are given in Table 6.1 (ICRP, 2007).

The total equivalent dose, H_T, is the sum of $H_{T,R}$ over all radiation types,

$$H_T = \sum H_{T,R} \quad \text{Unit : sievert, } 1\,\text{Sv} = 1\,\text{J kg}^{-1} \tag{6.1.2}$$

Table 6.1: ICRP recommended radiation weighting factors (ICRP, 2007).

Radiation type	Radiation weighting factor, w_R
Photons	1
Electrons and muons	1
Protons and charged pions	2
Alpha particles, fission fragments, heavy ions	20
Neutrons	A continuous function of neutron energy

Effective dose

Effective dose is defined as

$$E = \sum w_T H_T \quad \text{Unit : sievert, } 1\,\text{Sv} = 1\,\text{J kg}^{-1} \tag{6.1.3}$$

where w_T is the tissue weighting factor and H_T is the equivalent dose in a tissue or organ. Numerical values of w_T are given in Table 6.2

Summing all tissues in the body will give $\Sigma w_T = 1$

Committed equivalent dose

Committed equivalent dose is defined as

$$H_T(\tau) = \int_{t_0}^{t_0+\tau} \dot{H}_T(t)\,dt \tag{6.1.4}$$

Table 6.2: ICRP recommended tissue weighting factors (ICRP, 2007).

Tissue	Tissue weighting factor, w_T	$\sum w_T$
Bone marrow (red), colon, lungs stomach, breast, remainder tissues	0.12	0.72
Gonads	0.08	0.08
Bladder, oesophagus, liver, thyroid	0.04	0.16
Bone surface, brain, salivary glands, skin	0.01	0.04
Total		1.00

Remainder tissues: Adrenals, extrathoracic region, gall bladder, heart, kidneys, lymphatic nodes, muscle, oral mucosa, pancreas, prostate, small intestine, spleen, thymus, uterus/cervix.

where t_0 is the time of intake, $\dot{H}_T(t)$ is the equivalent dose rate at time t in organ or tissue T and τ is the time elapsed after time of intake.

If τ is not specified it is taken as 50 y for adults and 70 y for children.

Committed effective dose:
Committed effective dose is defined in a similar way as

$$E(\tau) = \sum w_T H_T(\tau) \tag{6.1.5}$$

Collective effective dose
The collective effective dose, due to individual effective dose values between E_1 and E_2 from a specified source within a specified time period ΔT, is defined as

$$S(E_1, E_2, \Delta T) = \int_{E_1}^{E_2} E[\frac{dN}{dE}]_{\Delta T} \, dE \tag{6.1.6}$$

It can be approximated as $S = \Sigma_i E_i N_i$ where E_i is the average effective dose for a subgroup i and N_i is the number of individuals in this subgroup. The unit of the collective effective dose is Joule per kilogram and the special name is person-sievert.

Annual limit on intake (ALI)
ALI is the value of I that satisfies the following inequality

$$I \sum w_T H_{50,T} \leq 0.02 \, \text{Sv} \tag{6.1.7}$$

$H_{50,T}$ is the committed equivalent dose for $\tau = 50$ y. This means that if a person has an intake of activity corresponding to an ALI for 50 y, the equivalent dose year 50 will not be larger than 0.02 Sv, which fulfills the ICRP recommendation.

Derived air concentration (DAC)

DAC is defined as the concentration in air that would result in an activity inhalation of an ALI

$$DAC = \frac{ALI}{2000 \cdot 1.2}\ \mathrm{Bq\,m^{-3}} \tag{6.1.8}$$

where it is assumed that a person (Reference Man) works 2000 h per year and inhales $1.2\,\mathrm{m^3}$ air per h.

6.1.2 Transport of radionuclides in the body

I. Single uptake and exponential excretion.
The activity in the body after a single intake and exponential excretion is given by

$$A(t) = A_0 e^{-\lambda_{\mathrm{eff}} t} \tag{6.1.9}$$

where A_0 is the initial uptake of activity, $\lambda_{\mathrm{eff}} = \lambda_{\mathrm{b}} + \lambda_{\mathrm{f}}$ (effective decay constant), λ_{b} is the biological decay constant and λ_{f} is the physical decay constant.

The relation can be rewritten including the half lives instead. Then

$$A(t) = A_0 e^{(-\ln 2 t / T_{\mathrm{eff}})} \tag{6.1.10}$$

$$T_{\mathrm{eff}} = \frac{T_{\mathrm{b}} T_{\mathrm{f}}}{T_{\mathrm{b}} + T_{\mathrm{f}}} \tag{6.1.11}$$

II. Single uptake and general excretion equation.
The variation of activity in the body after a single intake and with a general excretion equation is given by

$$\frac{\mathrm{d}A(t)}{\mathrm{d}t} = -E(t) - \lambda_{\mathrm{f}} A(t) \tag{6.1.12}$$

where $E(t)$ is the excreted activity per time unit at time t and $A(t)$ is the activity in the body after time, t.

The relative decrease in activity is then given by

$$\frac{\mathrm{d}R(t)}{\mathrm{d}t} = \frac{\mathrm{d}(\frac{A(t)}{A_0})}{\mathrm{d}t} = \frac{-E(t)}{A_0} - \lambda_{\mathrm{f}} \frac{A(t)}{A_0} = -Y(t) - \lambda_{\mathrm{f}} R(t) \tag{6.1.13}$$

where $R(t)$ is the fraction of initial activity, A_0, that is remaining in the tissue at time t and $Y(t)$ is the fraction of the initial activity that is excreted per time unit at time t.

$R(t)$ and $Y(t)$ do not include the physical decay and can thus be used for different radioactive isotopes of the same element.

III. Continuous contamination of the radionuclide.
The activity in the tissue at time t is given by the relation

$$A(t) = I \int_{0}^{t} R(t) e^{-\lambda_{\mathrm{f}} T}\,\mathrm{d}T \tag{6.1.14}$$

where I is the intake/time unit.

IV. Cumulated activity.
By integrating the activity $A(t)$ over time, the cumulated activity $\tilde{A}$ is obtained.

$$\tilde{A} = \int_0^t A(\tau)\,d\tau \qquad (6.1.15)$$

If $A(t)$ is decreasing exponentially with time then

$$\tilde{A} = \frac{A_0}{\lambda_{\text{eff}}}(1 - e^{-\lambda_{\text{eff}}t}) \qquad (6.1.16)$$

If $t \to \infty$ then

$$\tilde{A} = \frac{A_0}{\lambda_{\text{eff}}} = \frac{A_0 T_{\text{eff}}}{\ln 2} \qquad (6.1.17)$$

This approximation holds when t is much larger than the effective half life.

The absorbed dose rate and absorbed dose is obtained by multiplying the activity or the cumulated activity with a factor S Gy/(Bqs), giving the relation between absorbed dose rate and activity or absorbed dose and cumulated activity. The factor S is depending on type and energy of radiation, and the size and shape of the organ. This factor can also be used to calculate the absorbed dose due to activity in one organ to other organs, and then also the distance between the organs will be of interest. MIRD (Medical Internal Radiation Dose Committee) has calculated and tabulated data for S for different organs and human phantoms corresponding to different sizes, from babies to adults. These tables and other publications from MIRD can be downloaded from http://interactive.snm.org/.

6.1.3 Radiation shielding calculations

Radiation sources
Radiation sources are characterized by their strength and geometry. The simplest source is the point source and the other sources can be obtained by a summation of point sources. Normally one can divide the sources into the following categories:

Point source S (Bq)
Line source S_L (Bq m^{-1})
Area source S_A (Bq m^{-2})
Volume source S_V (Bq m^{-3})

Instead of defining the source strength in Bq, it is often defined as the number of particles, in our cases mainly photons, emitted per second. In the solutions discussed in this material, the definitions using Bq are used. Then it is important to multiply with the number photons per decay to obtain the fluence rate. As often photons of different energies are emitted, separate calculations have to be performed for each energy. The energy fluence rate is obtained by multiplying the fluence rate for each energy with the respective energy. This means that all calculations are a sum of calculations for the different energies.

Primary photons

Plane parallel beams. The attenuation of the fluence rate of a plane parallel beam passing through an absorber with thickness x and the linear attenuation coefficient μ is obtained through the relation

$$\dot{\Phi}(x) = \dot{\Phi}(0)e^{-\mu x} \tag{6.1.18}$$

This holds if only primary photons are included in the calculations. This can be assumed to hold for a narrow well collimated beam, both in front of and behind the absorber (see Fig. 6.1). This situation is sometimes called narrow beam or good geometry.

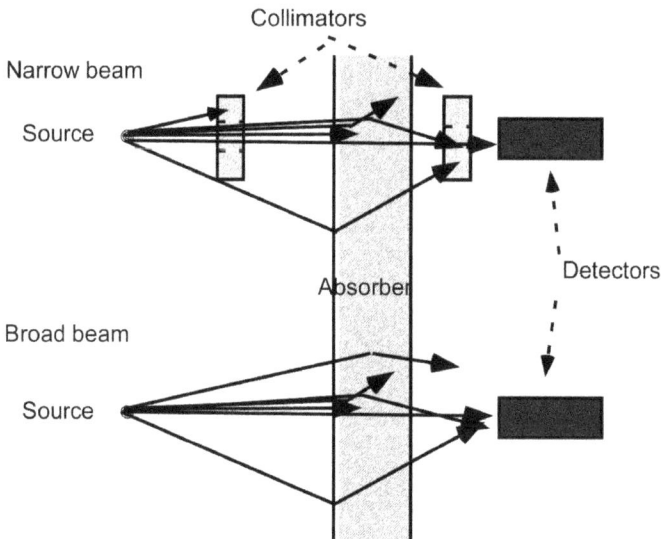

Figure 6.1: Transmission of photons trough a shielding material in "narrow" and " broad" beam conditions.

6.1.3.0.1 Point source.
With a point source there will, besides the attenuation, be a decrease of the fluence rate due to the divergence of the beam. Sometimes the quantity

$$\frac{e^{-\mu r}}{4\pi r^2} \tag{6.1.19}$$

is called the *point kernel*, as it describes the variation of the fluence rate with the distance from the source. As all sources can be assumed to consist of an infinite number of point sources, the fluence rate from any source can be calculated using the point kernels.

The fluence rate of the primary photons at a distance r from a point source with the source strength S (Bq), and f mono-energetic photons per decay, in an absorbing medium with the attenuation coefficient μ, is obtained through the relation

$$\dot{\Phi}(r) = \frac{Sf}{4\pi r^2} e^{-\mu r} \tag{6.1.20}$$

Secondary photons
Build-up factor. Normally secondary photons produced by interaction of the primary photons have to be included in the calculations (see Fig. 6.1). These secondary photons are mainly Compton photons, but to a smaller extent also annihilation photons and fluorescence x rays. Calculation of the secondary photon fluence rate for a certain geometry is complex and often simplifications are made. Often the primary fluence is multiplied with a build up factor, B, that is defined as

$$B = 1 + \frac{\text{contribution from secondary photons}}{\text{contribution from primary photons}} \tag{6.1.21}$$

This can for an isotropic, monoenergetic point source in a medium be written as

$$B_k(r, E_0) = \frac{\int_0^{E_0} f_k(E)\Phi(r, E)\,dE}{f_k(E_0)\Phi_0(r, E_0)} \tag{6.1.22}$$

The factor f_k defines which effect that the build-up factor is defined for. If f_k is equal to unity, then the buildup factor is related to the fluence. If it is equal to E, then it is related to the energy fluence and if it is equal to $\mu_{en}E$ it is related to the absorbed dose and so on. Φ_0 and Φ are the fluences of the primary and the total number of particles respectively.

The build up factor B is a function of several parameters:

a) Photon energy. B increases generally with decreasing photon energy in elements with low atomic numbers, where the Compton effect dominates down to low energies. For elements with a high atomic number, B will instead decrease at low

Table 6.3: Ratio between the build-up factor in a finite and an infinite medium. Data from Berger and Doggett (Berger and Doggett, 1956).

Material	Photon energy (MeV)	$1\mu x$	$4\mu x$	$16\mu x$
			$[B(\mu x, \mu x])-1]/[B(\mu x, \infty])-1]$	
Water	0.66	0.66	0.78	0.78
	1.0	0.72	0.82	0.83
	4.0	0.89	0.92	0.93
Lead	0.66	0.95	0.98	0.98
	1.0	0.98	0.99	0.99
	4.0	0.99	0.99	0.99

photon energies as the photo-electric effect, where most of the photon energy is transferred to the electron, will start to dominate.

b) <u>Atomic number of the absorber.</u> B decreases with increasing atomic number, as the probability for both the photo-electric effect and the pair production increases with increasing atomic number, giving rise to less secondary photons.

c) <u>Penetration depth.</u> B increases continuously with increasing value of the penetration depth, μx.

d) <u>Geometry.</u> Normally two geometries are discussed, a plane parallel infinite beam or a point source in an infinite medium. Most tables and equations thus give data for B in an infinite extension of the absorber. This generally does not correspond to the real geometry and B overestimates the contribution from secondary photons, mainly due to the backscattered photons, which should not be included in calculations of the fluence rate behind an absorber. The difference between the correct and the calculated fluence rate using these values of B is largest for low energies and low atomic numbers. The difference can be as large as up to 30%. Table 6.3 gives relations between the build up factor in a finite medium, $B(\mu x, \mu x)$ and an infinite medium $B(\mu x, \infty)$.

e) <u>Quantity studied.</u> The build-up factor is different depending on which quantity that is studied, fluence, energy fluence, absorbed dose and so on. The fluence build-up factor is larger than the buildup factor for the energy fluence as the energy of the secondary photons is lower than for the primary photons. It is thus important to use buildup factors for the quantity that is of interest.

Data for build-up factors can be found tabulated for some simple geometries in many textbooks and compilations on radiation shielding.

There are also some empirical equations where the parameters are fit to values calculated by analytical methods, that may be used for obtaining the buildup factor.

One example is the Taylor expression.

$$B(E_0, \mu r) = Ae^{-\alpha_1 \mu r} + (1 - A)e^{-\alpha_2 \mu r} \tag{6.1.23}$$

where A, α_1 and α_2 are parameters that vary with energy and medium.

Another commonly used expression is the Berger expression (Berger, 1956; Chilton, 1968).

$$B(E_0, \mu r) = 1 + a\mu r e^{b\mu r} \tag{6.1.24}$$

where the parameters a and b vary more slowly with energy than the parameters for the Taylor expression. Another advantage with this expression is that the contribution from secondary particles is separate.

Data for the parameters for the Berger expression are e.g. found in Tables A4:9 and A4:10 in Chilton et al (Chilton et al, 1984).

In some situations it is more practical to use experimental obtained transmission data. This is often the case when determining the shielding in diagnostic radiology and in therapeutic treatment rooms, where the approximation of infinite beams is in-appropriate.

When more accurate calculations are needed the best method is to use the Monte Carlo method, where the real irradiation geometry can be simulated. These calcula-tions may however be time consuming and in radiation protection a high accuracy is not always needed.

6.1.3.0.2 Shielding including secondary photons.

When including the secondary photons, the equation to calculate the fluence rate, is for a plane parallel beam given by

$$\dot{\Phi}(x) = \dot{\Phi}(0)B(\mu x)e^{-\mu x} \tag{6.1.25}$$

The corresponding equation for a point source is given by

$$\dot{\Phi}(r) = \frac{SfB(\mu r)}{4\pi r^2} e^{-\mu r} \tag{6.1.26}$$

Thus the total fluence is simply obtained by multiplying the relation for the primary fluence with the buildup factor B. Note that the buildup factor takes into considera-tion the different energies of the secondary and primary photons if needed. It is, as already mentioned, important to use the correct factors depending on the quantity one is interested in.

6.1.3.0.3 Reflexion coefficient of radiation, Albedo.

In many situations the backscattered radiation is of interest. If a medium has the shape of an infinite slab and radiation is incident on one of its sides, the reflexion coefficient or the Albedo factor is defined as the ratio of the amount of radiation reflected from the slab to the amount of radiation incident on the slab. The definition assumes that the reflected radiation is emitted from the same point as where it entered the slab (See Fig. 6.2), even if the reflexion is not a surface effect, but the interactions are at different depths in the slab.

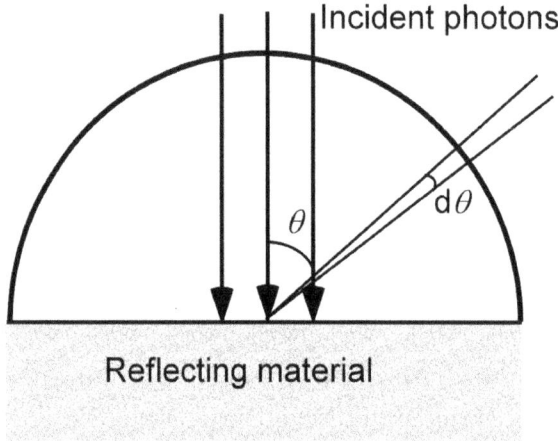

Figure 6.2: Reflexion of photons from a surface.

The reflexion coefficient for the number of particles may be defined as

$$R_N = 2\pi \int_0^{\pi/2} \int_0^{E_0} N(\theta, E) \sin\theta \, d\theta \, dE \qquad (6.1.27)$$

where $(N(\theta, E))$ is the relative number of particles reflected differential in angle and energy.

Extended sources

Fluence rate from line sources. The fluence rate for primary photons from a line source at a point P (See Fig. 6.3) and a total attenuation thickness of $\Sigma\mu_i x_i$ is given by

$$\dot{\Phi}_{P,L} = \frac{S_L f}{4\pi h} [F(\theta_2, \Sigma\mu_i x_i) + F(\theta_1, \Sigma\mu_i x_i)] \qquad (6.1.28)$$

where F is the Sievert integral defined as

$$F(\Theta, \mu x) = \int_0^\theta e^{-\mu x / \cos\theta} \, d\theta \qquad (6.1.29)$$

Table 6.4 gives data for $F(\theta, \Sigma\mu_i x_i)$.

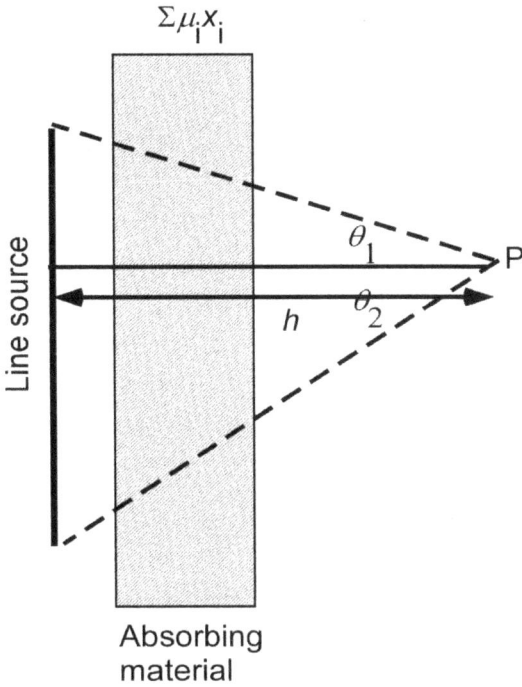

Figure 6.3: Line source with an absorbing material.

Fluence rate from area sources. The fluence rate of primary photons from a plane circular area to a point P (See Fig. 6.4), centrally positioned above the area at a height h and with a total attenuation thickness of $\Sigma\mu_i x_i$ is given by

$$\dot{\Phi}_{P,A} = \frac{S_A f}{2}[E_1(\Sigma\mu_i x_i) - E_1(\theta, \Sigma\mu_i x_i \sec\theta)] \tag{6.1.30}$$

When the radius of the area goes to infinity the second factor goes to zero and the relation becomes

$$\dot{\Phi}_{P,A} = \frac{S_A f}{2}E_1(\Sigma\mu_i x_i) \tag{6.1.31}$$

$E_1(\Sigma\mu_i x_i)$ is the exponential integral of the first order obtained from the general relation

$$E_n(x) = x^{n-1}\int_{x}^{\infty}\frac{e^{-y}\,dy}{y^n} \tag{6.1.32}$$

Observe that for an infinite area, the height over the surface is not included in the equation, and it is only the attenuation thickness that is of interest. For areas not circular but more or less symmetric it is possible to approximate the real area with a circular area of the same size. Table 6.5 gives values of the exponential integral functions $E_1(x)$ and $E_2(x)$.

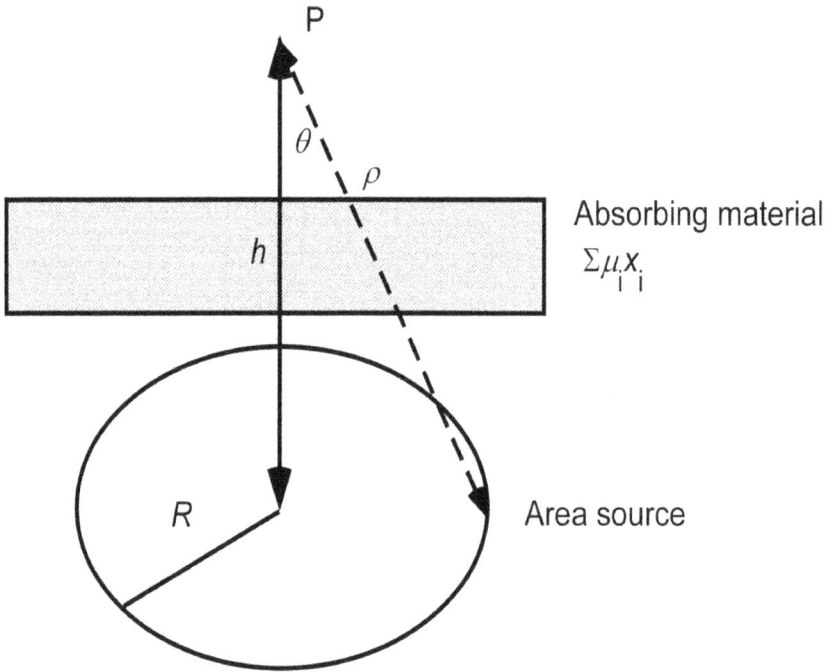

Figure 6.4: Area source with absorbing material.

Fluence rate from volume sources. The fluence rate at a point P from a semi-infinite volume source with a thickness d (See Fig. 6.5), attenuation coefficient μ_s and an absorbing material with an attenuation thickness $\Sigma \mu_i x_i$ is obtained by dividing the volume source into several thin slices and then integrate the fluence from each slice. The fluence rate of primary photons is then obtained as

$$\dot{\Phi}_{P,V} = \frac{S_V f}{2\mu_s}[E_2(\Sigma \mu_i x_i) - E_2(\Sigma \mu_i x_i + \mu_s d)] \tag{6.1.33}$$

If the volume source can be assumed to have an infinite thickness then the equation is simplified to

$$\dot{\Phi}_{P,V} = \frac{S_V f}{2\mu_s}E_2(\Sigma \mu_i x_i) \tag{6.1.34}$$

If instead there is no absorbing material between the source and the calculation point P the equation becomes

$$\dot{\Phi}_{P,V} = \frac{S_V f}{2\mu_s}[1 - E_2(\Sigma \mu_s d)] \tag{6.1.35}$$

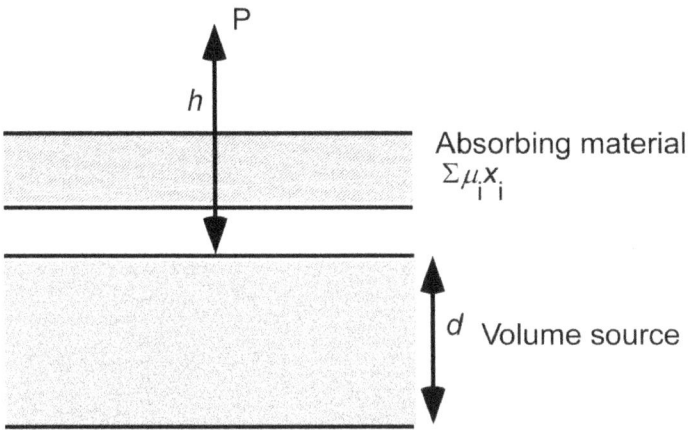

Figure 6.5: Volume source with absorbing material.

6.1.3.1 Radiation shielding at radiotherapy installations

When installing a high energy radiation treatment facility, the protection of both patients and staff need to be considered. In this section only the determination of the
shielding of the walls of the treatment room will be discussed. For further discussion
see e.g. ICRP 33 (ICRP, 1982) and NCRP 151 (NCRP, 2005). Three main components have
to be included in the discussion, primary radiation, leakage radiation and scattered
radiation. Production of neutrons and induced activity will not be considered here. In
the discussion it is important to consider the type of areas that shall be protected, as
the acceptable permissible dose depends on the group of people to be protected (i.e.
the general public or staff). The national recommendations should be followed here.
Typical values could be a weekly dose equivalent of 0.02 mSv/week for uncontrolled
areas and 0.1 mSv/week for controlled areas NCRP 151 (NCRP, 2005).

Most accelerator installations are isocentric. This will imply that two walls will
not be hit by the primary beam, but only by leakage and scattered radiation. Thus the
thickness of these walls may be reduced. Determination of the protective barriers for
these three components will be discussed separately below.

Primary radiation. When determining the necessary shielding for walls which
are directly hit by the beam the relation

$$B_\mathrm{P} = \frac{Pd^2}{WUT} \tag{6.1.36}$$

is sometimes proposed as a starting point (ICRP, 1982). B is the transmission through
the barrier to be determined, P is the is the acceptable equivalent dose, d is the distance from the source to the location of interest, W is the work load (absorbed dose at
1 m), U is the use factor (fraction of the treatments directed to the wall of interest), T
is the occupancy factor (taking into consideration of how much the space behind the
barrier is used). The calculation time can be either per week or per year. It is important

when deciding of the different factors, to realize that they may change in the future due to changes both in the use of the building and to changes in treatment methods. Thus it is important to use conservative assumptions and in practice the doses are often far below the regulations.

To determine the necessary wall thickness for a certain B, either half-value thickness (HVT), tenth-value thickness (TVT) or experimental transmission curves may be used (ICRP, 1982) as these beams have to be considered as broad beams.

Scattered radiation. In a similar way the relation for scattered radiation can be expressed as

$$B_S = \frac{P d_s^2}{W T S} \tag{6.1.37}$$

P and T are the same and W is the same unless the distance between source and scatterer is not 1 m. d_s is the distance from scattering object (patient) to the location of interest. S is the fraction of incident absorbed dose rate scattered to 1 m. When determining barrier thickness for secondary radiation it is important to remember that the scattered radiation has a lower energy than the primary one. Independent of the energy of the primary radiation, the energy of the radiation scattered more than $90°$ is lower than around 0.5 MeV. Thus the TVT and transmission curves for scattered radiation shall be used. In many installations the scattered radiation gives a small contribution to the total dose and can be neglected in barrier determinations.

Leakage radiation. For leakage radiation, transmission curves are not recommended to be used but only tenth-value thicknesses. The number of tenth-value thicknesses N_{TVT} is given by

$$N_{TVT} = \log_{10} \frac{W_L T}{d^2 P} \tag{6.1.38}$$

W_L is the leakage absorbed dose at 1 m from the source. For high energy x rays this value is at 1 m from the target often 0.1 % of the primary absorbed dose at 1 m (NCRP, 2005).

A wall is normally hit by both scattered radiation and leakage radiation. If the difference is larger than 1 TVT, which is the most common situation, the thicker shield thickness should be used. If the difference is smaller than 1 TVT, then 1 HVT should be added to the thicker shield.

6.2 Exercises in Radiation Protection

6.2.1 Radioecology

Exercise 6.1. A company is using high activities of ^{57}Co. The waste activity is first released through waste water into a water basin and then, after passing through a filter, out into the sewage. The renewal of the water into the basin is 5.0 m^3 per day. The volume of the basin is 200 m^3. The contaminated water is assumed to immediately mix with the water in the basin. A check of the activity concentration in the water in the basin after 200 days shows that it is 2.50 kBq m^{-3}. This was regarded as too much and actions were made to reduce the release of the activity into the basin with a factor of 3.0. Calculate the activity concentration in the basin 100 days later.

Exercise 6.2. A laboratory can measure an activity of 0.20 Bq ^{89}Sr in a urine sample. Assume that workers handling ^{89}Sr, have a daily urine sample measured every 30 days. Calculate the lowest detectable intake in the "worst situation" i.e. when the worker was exposed to the contamination 30 days before the sample was taken. Assume that 30% of the activity is excreted through the urine and that the excretion equation is given by

$$Y(t) = 0.12 \cdot e^{-\ln 2 \cdot t / 2.4} + 0.08 \cdot t^{-1.2} \text{ (t in days)}$$

Exercise 6.3. A person is working in an environment in which the concentration of activity of ^{3}H is 0.10 DAC (derived air concentration). He is working in this environment for 180 days. Calculate the expected total effective dose he will obtain. ^{3}H has a biological half life of 12.0 days.

Exercise 6.4. You are responsible for a laboratory using ^{210}Po. There is a detector available that can measure the activity of ^{210}Po in the urine. The detection limit is 15 Bq ^{210}Po in the daily urine. You are interested in measuring a single uptake of 20.0 kBq ^{210}Po. How often is it necessary to perform urine measurements to be able to detect such an uptake? The retention equation for ^{210}Po in the kidneys is

$$R(t) = e^{-\ln 2 \cdot t / 40} \text{ (t in days)}$$

10% of the activity is taken up by the kidneys and excreted through the urine.

Exercise 6.5. A person works for 60 days in an area with a high concentration of tritium in the drinking water. A continuous stay in this environment would result in an activity intake corresponding to one ALI, calculated for an effective dose of 20 mSv per year. Calculate the effective dose during the 60 days the person is staying in this environment. Calculate also the total effective dose integrated over infinite time. The biological half life is 10 days.

Exercise 6.6. A research institute is using ^{3}H. By mistake the waste water became contaminated. This waste water is used by cattle drinking 70 dm^3 of the water every day. The activity concentration of the water is 390 kBq dm^{-3}. This is going on for 40 days until the contamination is discovered. The cattle will then drink uncontaminated water. Calculate the total absorbed dose to an animal with a mass of 400 kg (assuming infinite time). The biological half life is obtained by assuming that the water intake and excretion is the same and the water content is 260 kg. When calculating the mean absorbed dose assume that the activity is distributed uniformly in the body.

Exercise 6.7. The chemical toxicity for nickel carbonyl, $Ni(CO)_4$, gives the limit for the atmospheric concentration to $1.0 \cdot 10^{-10}\%$ per mass unit. In an experiment nickel carbonyl, corresponding to the chemical limit, tagged with 100% ^{63}Ni is used. Is this in agreement with the radiological limit expressed in ALI for an effective dose of 20 mSv per year, i.e. the exposed persons are assumed to be radiological workers? The whole body is supposed to be the critical organ (70 kg). Assume that the breathing capacity is $10.0 \text{ m}^3 \text{ d}^{-1}$ and that 50% of the inhaled activity is absorbed. The biological half life is 2.0 years.

Exercise 6.8. When estimating risks from the food contaminated with radioactivity from fallout, it is necessary to consider how often a certain foodstuff is consumed. Compare the yearly emitted electron- and photon energy from the radioactivity in the body for the two following situations:
a) A single intake of 0.60 kg of lobster from Sellafield. The lobster contains 13000 Bq kg^{-1} of ^{137}Cs.
b) A "continuous" daily intake of 0.50 dm^3 milk, with the concentration 100 Bq dm^{-3} of ^{137}Cs. The excretion of Cs from the body can be approximated with a biological half life of 100 days.

Exercise 6.9. After the reactor power accident in Chernobyl the maximal concentration of ^{137}Cs in food was in Sweden limited to 300 Bq kg^{-1}. The aim was that the population during their life shall not obtain an activity content higher than 30 000 Bq per person. Calculate the maximum mass intake of the contaminated food per day if this is to be fulfilled assuming constant activity concentration. The average length of life in Sweden is supposed to be 80 years. The retention function for ^{137}Cs is given by

$$R(t) = 0.10 \cdot e^{-0.347t} + 0.90 \cdot e^{-0.00630t} \text{ (t in days)} \tag{6.2.1}$$

Exercise 6.10. A pond is by mistake contaminated with ^{137}Cs. A test measurement shows that the concentration of the activity is 2.5 kBq dm^{-3} water. The pond has an exchange of water of 1000 dm^3 per day and the volume of the pond is 100 000 dm^3 water. The mixture of the water is momentarily so the concentration of activity in the pond is uniform. The water in the pond is drunk by cows on a pasture surrounding the pond. Calculate the maximum activity in the cows if they drink 70 dm^3 water per day. The half life for Cs in the cow is assumed to be 100 days.

Exercise 6.11. One of the problems when using fusion reactors is the production of tritium. In a river the concentration of ^{3}H in the water is 800 Bq dm^{-3}. Calculate the effective dose per year for a continuous intake of water, assuming that a person drinks 2.0 dm^3 water per day. Hydrogen is in this calculation approximated to be evenly distributed in the body with a mass of 70 kg. The biological half life for hydrogen is 10 days.

Exercise 6.12. You are working at a hospital, that is using ^{131}I for treatment of diseases in the thyroid. There is an automatic system for distribution of the activity to the patient. At some occasion it is discovered that a tube in the system is broken and some activity is leaking out into a room with a volume of 20 m^3. The system is assumed to be continuously leaking ^{131}I with a velocity of 0.20 MBq h^{-1}. The activity is mixed momentarily with the air and the concentration is approximated to be evenly distributed in the room. The ventilation exchange rate in the room is 0.50 h^{-1}. A laboratory assistant has been working in the laboratory the whole day, i.e. 8.0 h. Assume that the leakage started already in the morning, which means that there has been a leakage the whole day. Calculate the activity in the thyroid of the laboratory assistant at the end of the day, if 30% of the inhaled activity is taken up in the thyroid and excreted with a biological half life of 75 days. The inhaling rate is 0.020 m^3 min^{-1}.

6.2.2 Point Radioactive Sources

Exercise 6.13. A ^{60}Co source is placed in a safety box in a storage room. In another room close to the storage, a person is sitting all the day, i.e. 40 h per week. Calculate the absorbed dose to the person during one week, using the following data. The activity of the source is 0.63 GBq. The wall of the safety box is made of 30 mm iron. The wall between the rooms is made of 15.0 cm concrete (ρ = 2.35 · 10^3 kg m^{-3}). The absorbed dose is calculated to a small mass of water "free in air" at a distance of 60.0 cm from the source. The build-up factor due to the secondary photons is 3.0. Is it acceptable that the person is sitting there or is it necessary to improve the shielding?

Exercise 6.14. In a student's laboratory radioactive sources are stored in a safe. The dose rate (to a small mass of water) outside the safe should be less than 10 μGy h^{-1}. Assume that the dominating radioactive source in the safe is ^{60}Co. Which is the highest possible activity of this source in order to fulfill the radiation protection requirements? Assume the following geometry: The shortest distance from the radioactive source to the outside of the safe is 130 mm. The wall of the safe is made of 20 mm iron. For calculation of the build-up factor use the Berger expression with the parameters a=0.955 and b=0.024.

Exercise 6.15 To decrease the absorbed dose when working with radioactive sources, it is possible to work behind a shield made of lead. Sometimes this however prolongs the working time and the absorbed dose can be higher than without the shield. Which of the following working situations gives the lowest absorbed dose?

Method 1: The radioactive source (^{60}Co) is positioned 50 cm from the person behind a 40 mm thick lead shield. The working moment takes 600 s.

Method 2: The radioactive source (^{60}Co) is positioned at a distance of 40 cm from the person without any lead shield. The working moment now takes 140 s.

For calculation of the build-up factor use the Berger expression with the parameters a=0.33 and b=-0.011. If the activity of the source is 200 GBq, calculate the kerma rate to air at the calculation point where the person is situated. The mean energy of the photons may be used.

Exercise 6.16. A young patient is to be treated with ^{60}Co-γ-radiation and the father has then to be in the treatment room during the irradiation. The kerma rate to air is 21.8 mGy s^{-1} at a distance of 1.00 m from the source. SSD is equal to 0.80 m and the treatment time is 150 s every time. The distance to the father is 4.00 m. He is only hit by 90° scattered radiation from the patient and leakage radiation from the treatment head. The thickness of the patient is 20 cm. The treatments are made 5 days per week and during 4 weeks.

The father is positioned behind a lead shield. How thick must this lead shield be if the father shall not get more than a total effective dose of 100 μSv? The kerma rate of the scattered radiation 1.00 m from the scatterer (the patient) is 0.10% of the kerma rate of the primary beam at position of the center of the scatterer. Use transmission curves for broad beams to calculate the necessary shielding thickness. The leakage dose rate at 1.0 m is assumed to be 0.1% of the primary dose rate at the isocenter. Tenth value thickness in lead is 4.0 cm (ICRP, 1982). Conversion factor effective dose to air kerma = 1.09.

Exercise 6.17. A 6 MV linear accelerator will be installed in a room previously used for a ^{60}Co radiotherapy unit. The wall thickness will then need to be increased. The treatment room is outlined in Fig. 6.6 where the lines indicate the outer contour of the walls. At point A there is a control room and point B is situated outside the building, where it is unlikely that anybody will be there for a long time. According to national regulations, considering typical values for work load, use and occupancy factors, this implies that the absorbed dose rate at A shall be <10 μGy h^{-1} and at B <100 μGy h^{-1}.

The linear accelerator is mounted isotropically and the primary beam can hit A, but not B. The maximum absorbed dose rate at the isocenter (1.00 m) is 4.0 Gy min^{-1}. The leakage dose rate at 1.0 m is assumed to be 0.1% of the primary dose rate at the isocenter. The scattered radiation 1.0 m from the isocenter is 0.03% of the primary dose rate at that point. For the calculation of the wall thicknesses, data from ICRP 33 (ICRP 1982) or similar compilations can be used. Calculate the necessary wall thicknesses for the walls at A and B.

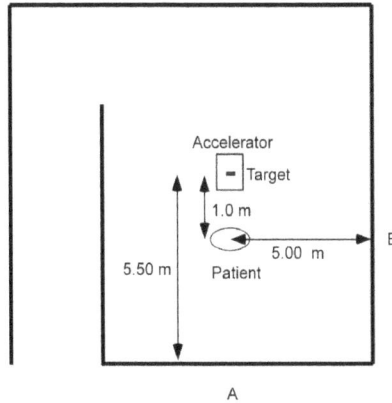

Figure 6.6: Outline of a radiotherapy treatment room with a linear accelerator.

Exercise 6.18. At a hospital, a room shall be used for brachytherapy patients, i.e. patients with intracavity radioactive sources. It is important that patients in a nearby room shall not receive absorbed doses which are too high. Calculate the necessary thickness for the wall between the rooms. See Fig. 6.7. Assume that there are patients with radioactive sources for 50 h per week. In the nearby room the same patient can stay for a week. The effective dose to the patient in the nearby room shall be lower than 0.020 mSv. The dose may be calculated as the kerma to a small mass of water "free in air" at the center of the patient, without including any absorption in the patient.

The brachytherapy sources are ^{137}Cs with an activity of 2.60 GBq. The wall is made of concrete with a density of $2.35 \cdot 10^3$ kg m^{-3}. Each patient has only one source and only one patient is in the room. The self absorption in the brachytherapy patient is 30%. The air kerma constant for ^{137}Cs=$20.3 \cdot 10^{-18}$ Gy s^{-1} Bq^{-1} m^2.

6.2.3 Extended Radioactive Sources

Exercise 6.19. In gynecological radiotherapy, radioactive sources which are inserted into the body, are used. Such a radiation source may consist of a line source inside a material which protects the source from the tissue, but also absorbs all electrons and low energetic photons that are emitted from the source. Such a radiation source has dimensions as in Fig. 6.8, where the cover made of platinum (Pt) has a thickness of

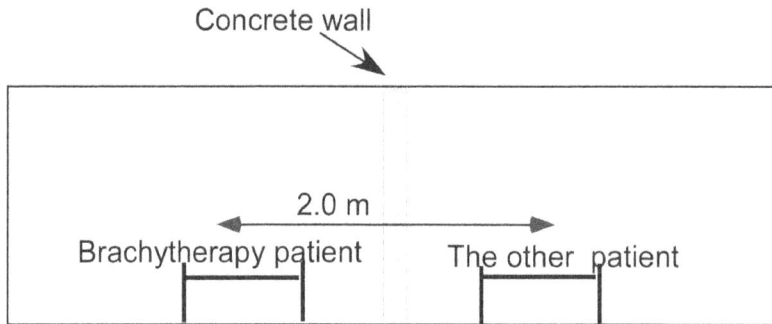

Figure 6.7: Brachytherapy patient in a room next to another patient.

0.65 mm. The source is placed in water. Which time is needed to obtain an absorbed dose at P of 2.0 Gy, if the activity of the source is 57.5 GBq? Only primary photons have to be included in the calculations.

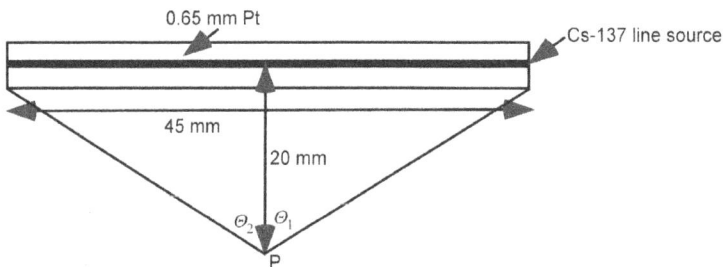

Figure 6.8: ^{137}Cs brachytherapy source with a Pt-cover.

Exercise 6.20. A betatron (a circular electron accelerator), may be radioactively contaminated trough gamma-n and electron-n reactions when high accelerating energies are used. Assume that the betatron may be regarded as a circular radioactive source with the diameter 1.50 m. Calculate the kerma to a small mass of water at the center of the betatron during the first 30 min after the betatron has been switched off. The important radionuclide is ^{15}O and the line source activity at the time of switch off is 1.30 GBq m^{-1}. Between the source and the calculation point is the wall of the betatron that is made of iron and has a thickness of 10.0 mm. Only primary photons have to be included in the calculations.

Exercise 6.21. In a room there is a tube in which a radioactive gas is passing (See Fig. 6.9). The closest distance from the tube to a work place is 2.00 m. The line source activity in the tube is $1.20\,\mathrm{GBq\,m^{-1}}$ and consists of $^{18}\mathrm{F}$. Calculate the kerma rate to air at the work place. The tube wall is made of iron and has a thickness of 2.0 mm. Assume that the positrons are annihilated directly at the inside of the tube and that the contribution from secondary photons can be neglected. The activity can be regarded as a line source and the contribution from the activity in the tube outside the room can be neglected. Is this work place acceptable regarding radiation protection recommendations?

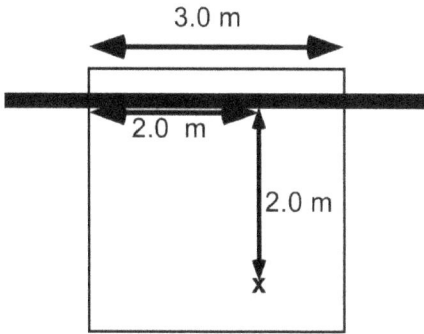

Figure 6.9: Room with a tube containing radioactive gas and a work place situated 2.0 m from the tube.

Exercise 6.22. In an experiment, cows on a pasture contaminated with $^{131}\mathrm{I}$ are irradiated. The concentration of activity is $450\,\mathrm{MBq\,m^{-2}}$. The pasture is large so an infinite extension may be assumed in the calculations. Calculate the kerma to tissue at a point 1.50 m over the ground, if the cows are on the pasture for 14 days. The air pressure is 100 kPa and the mean air temperature 15 °C. Include only the most frequent photon energy in the calculations.

Exercise 6.23. After the Chernobyl accident, parts of Sweden were contaminated with radioactivity. The area around Gävle was heavily contaminated, and an area source activity up to $80\,000\,\mathrm{Bq\,m^{-2}}$ was measured. The main radionuclide was $^{137}\mathrm{Cs}$. Calculate the maximal kerma rate to air 1.5 m above the ground for a circular area with the radius 200 m with this concentration. The air pressure is 101.3 kPa and the air temperature 20 °C. The contribution from the characteristic x rays and the production of secondary photons can be neglected.

Exercise 6.24. During winter Sweden is typically covered with snow. The snow will decrease the natural background radiation from the ground. Calculate the decrease in air kerma rate 1.50 m above the ground, if the ground can be approximated with an infinite plane surface source with the activity $10000\,\mathrm{Bq\,m^{-2}}$. Assume that 0.80 photons with the energy 0.80 MeV are emitted per disintegration. The snow thickness is 0.50 m and the density is $0.25\,\mathrm{kg\,dm^{-3}}$. The build-up factor is calculated using the formula according to Berger with the parameters a=1.74 and b=0.045. The same formula constants may be used for both air and snow (water). When calculating the kerma rate the same energy may be used for the primary and the secondary photons.

Exercise 6.25. Calculate the primary photon fluence rate at the water surface above a nuclear fuel rod, that is placed vertically, with its upper endpoint 2.00 m below the water surface.

The rod consists of uranium dioxide and is 4.00 m long, and may be regarded as a thin line source, but where the attenuation in the source should be considered along the length of the source. The activity in the source is $37 \cdot 10^{15}$ Bq and a photon with the energy 1.0 MeV is emitted per disintegration.

Exercise 6.26. A measuring room for low-activity measurements were built with a 0.40 m thick concrete wall, with the volume source activity $7.4\,\mathrm{kB\,m^{-3}}$. The radioactive nuclides decay by emitting two photons per disintegration with the average energy of 1.0 MeV. To reduce the background contribution from the wall it was covered with 20 mm lead.

Calculate the fluence rate 1.0 m from the wall, with and without the lead. The wall may be assumed to be semi-infinite. Estimate what error this approximation introduces if the wall has the dimension $6.0 \times 2.0\,\mathrm{m^2}$. The build-up factor is calculated using the formula according to Berger with the parameters for concrete a=1.27 and b=0.032 and for lead a=0.30 and b=-0.015.

6.3 Solutions in Radiation Protection

6.3.1 Radioecology

Solution exercise 6.1.
The change of the concentration of activity in the basin is given by the equation

$$\frac{\mathrm{d}C}{\mathrm{d}t} = U - \lambda_{\mathrm{eff}} C(t) \tag{6.3.1}$$

Solving the equation gives

$$C(t) = \frac{U}{\lambda_{\mathrm{eff}}}(1 - e^{-\lambda_{\mathrm{eff}}t}) \tag{6.3.2}$$

U is the input of activity to the basin per time and volume unit and $\lambda_{\mathrm{eff}} = \lambda_{\mathrm{f}} + \lambda_{\mathrm{b}}$ (effective decay constant)

Data:

$C(200\,\text{d})=2.5\,\text{kBq m}^{-3}$ (activity concentration in the basin after 200 d)

$\lambda_f = \ln2/270.9\,\text{d}^{-1}$ (physical decay constant)

$\lambda_b=5.0/200\,\text{d}^{-1}$ (rate of exchange of water in the basin)

$\lambda_{eff}=2.756\cdot10^{-2}\,\text{d}^{-1}$

Data inserted in the equation gives

$$2.5\cdot 10^3 = \frac{U}{0.02756}(1 - e^{-0.02756\cdot200})$$

$U=68.62\,\text{Bq m}^{-3}\,\text{d}^{-1}$

The concentration of activity 100 d after the decrease of the activity is given by the new activity rate released, which is $68.62/3\,\text{Bq}\,(\text{m}^{-3}\,\text{d}^{-1})$ and the decrease of the activity in the basin after 200 d, when the change in activity was introduced.

$$C = \frac{68.62}{3\cdot0.02756}(1 - e^{-0.02756\cdot100}) + 2.5\cdot 10^3 \cdot e^{-0.02756\cdot100} = 936\,\text{Bq m}^{-3}$$

Answer: The concentration of activity 100 d after decrease of release of the activity is $0.94\,\text{kBq m}^{-3}$.

Solution exercise 6.2.

Excretion of ^{89}Sr is given by the equation

$$Y(t) = 0.12 \cdot e^{-\ln 2\cdot t/2.4} + 0.08 \cdot t^{-1.2} \tag{6.3.3}$$

The excreted activity per day at time t is then given by

$$\frac{dA(t)}{dt} = A_0 fY(t)e^{-\ln 2\cdot t/T_{1/2}} \tag{6.3.4}$$

where

$T_{1/2}=50.5\,\text{d}$ (physical half life of ^{89}Sr)

$dA/dt=0.20\,\text{Bq d}^{-1}$ (minimal measurable excreted activity in the urine per day)

$f=0.30$ (fraction excreted through urine)

$t=30\,\text{d}$ (time between two measurements)

Data inserted in Eq. 6.3.4 gives

$$0.20 = A_0 \cdot 0.30 \cdot (0.12 \cdot e^{-\ln 2\cdot30/2.4} + 0.08 \cdot 30^{-1.2}) \cdot e^{-\ln 2\cdot30/50.5}$$

This gives

A_0=734 Bq

Answer: The minimum detectable intake is 0.73 kBq ^{89}Sr.

Solution exercise 6.3.
The activity in the body at time t with continuous intake is given by

$$A(t) = \frac{I}{\lambda_{\text{eff}}}(1 - e^{-\lambda_{\text{eff}}t}) \qquad (6.3.5)$$

After a time corresponding to several half lives, equilibrium is obtained and the activity in the body is

$$A(\infty) = \frac{I}{\lambda_{\text{eff}}}$$

where I is the intake/time unit and λ_{eff} is the effective decay constant.

In this situation an air activity concentration corresponding to a DAC (derived air concentration) will give an effective dose, E, during a year corresponding to 20 mSv. This gives the relation

$$E = \frac{I}{\lambda_{\text{eff}}}ST \qquad (6.3.6)$$

where S is the factor giving the relation between activity and effective dose rate and T is the time for dose calculation (one year).

Thus

$$I = \frac{E\lambda_{\text{eff}}}{ST} \qquad (6.3.7)$$

This relation is inserted in Eq. (6.3.5) to obtain the activity at time T_1.

$$A_1(T_1) = \frac{E\lambda_{\text{eff}}}{ST\lambda_{\text{eff}}}(1 - e^{-\lambda_{\text{eff}}T_1}) \qquad (6.3.8)$$

The activity then decays and the variation of activity with time after T_1 is given by

$$A_2(t) = \frac{E}{ST}(1 - e^{-\lambda_{\text{eff}}\cdot T_1})e^{-\lambda_{\text{eff}}\cdot t} \qquad (6.3.9)$$

The effective dose can then be obtained through Eq. (6.3.8) and Eq. (6.3.9).

The first 180 d:

$$E_1 = S\int_0^{T_1} \frac{E}{ST}(1 - e^{-\lambda_{\text{eff}}t})\,dt = \frac{E}{T}\left[T_1 + \frac{e^{-\lambda_{\text{eff}}T_1}}{\lambda_{\text{eff}}} + \frac{1}{\lambda_{\text{eff}}}\right] \qquad (6.3.10)$$

Data:
$\lambda_{\text{eff}} = \lambda_{\text{b}} = \ln2/12 = 0.05776\,\text{d}^{-1}$ as λ_{f} can be neglected.
E=2.0 mSv

$T=1.0$ y$=365$ d
$T_1=180$ d

Data inserted gives

$$E_1 = \frac{2.0 \cdot 10^{-3}}{365}[180 + \frac{e^{-0.05776 \cdot 180}}{0.05776} + \frac{1}{0.05776}] = 0.891 \cdot 10^{-3} \text{ Sv}$$

After 180 d

$$E_2 = \frac{2.0 \cdot 10^{-3} \cdot S}{S \cdot 365}(1 - e^{-0.05776 \cdot 180}) \int_0^\infty e^{-0.05776 \cdot t} \, dt = \frac{2.0 \cdot 10^{-3}}{365} \frac{(1 - e^{-180 \cdot 0.05776})}{0.05776}$$

$E_2 = 0.095 \cdot 10^{-3}$ Sv.

The total effective dose is then $E=0.891+0.095=0.988$ mSv

Answer: The total effective dose is 0.99 mSv.

Solution exercise 6.4.
The variation of ^{210}Po activity in the body is given by the retention function corrected with the physical half life.

$$A(t) = A_0 e^{-\lambda_{\text{eff}} t} \tag{6.3.11}$$

$\lambda_{\text{eff}} = \lambda_f + \lambda_b$

The excretion is given by differentiation of Eq. (6.3.11)

$$\frac{dA}{dt} = -A_0 \lambda_{\text{eff}} e^{-\lambda_{\text{eff}} t} \tag{6.3.12}$$

where
$A_0=20.0$ kBq (initial activity of ^{210}Po)
$\lambda_f=\ln2/138.4$ d^{-1} (physical decay constant)
$\lambda_b=\ln2/40$ d^{-1} (biological decay constant)
$\lambda_{\text{eff}}=2.234 \cdot 10^{-2}$ d^{-1} (effective decay constant)

The minimum measured activity excreted in the daily urine should be 15 Bq. 10% of the activity is excreted through urine. Thus data inserted in Eq. (6.3.12) gives

$$15 = 0.1 \cdot 20 \cdot 10^3 \cdot 2.234 \cdot 10^{-2} \cdot e^{-2.234 \cdot 10^{-2} \cdot t}$$

Solving t gives $t=49$ d

Answer: It is necessary to measure at least every 49th day.

Solution exercise 6.5.

ALI (annual limit of intake) shall be calculated over a time span of 50 years. However, with the short biological half life in this case (T_b=10 d) it is possible to integrate over infinity.

The effective dose is given by Eq. 6.3.13 if the disintegrations are integrated over infinite time

$$E = \frac{ALI}{\lambda_{\text{eff}}} S \qquad (6.3.13)$$

Solving S (Sv Bq^{-1} s^{-1}) gives

$$S = \frac{E\lambda_{\text{eff}}}{ALI} \qquad (6.3.14)$$

Assume continuous intake, then

$$I = ALI/365 \text{ Bq d}^{-1}$$

The activity in the body after a time t is then given by

$$A(t) = \frac{I}{\lambda_{\text{eff}}}(1 - e^{-\lambda_{\text{eff}}t}) \qquad (6.3.15)$$

Integration over a time T, during which there is an intake, gives the number of disintegrations during this time

$$N(T) = \int_0^T A(t)dt = \frac{I}{\lambda_{\text{eff}}}(T + \frac{e^{-\lambda_{\text{eff}}T}}{\lambda_{\text{eff}}} - \frac{1}{\lambda_{\text{eff}}}) \qquad (6.3.16)$$

The number of disintegrations after the time T, where there is no intake of activity, is given by

$$N_\infty = \frac{A(T)}{\lambda_{\text{eff}}} = \frac{I}{(\lambda_{\text{eff}})^2}(1 - e^{-\lambda_{\text{eff}}T}) \qquad (6.3.17)$$

Data:
$A(T)$=activity at time T
E=20 mSv (permissible effective dose per year for a person in radiological work)
λ_f=ln2/12.3 y^{-1} (physical decay constant)
λ_b=ln2/10 d^{-1} (biological decay constant)
$\lambda_{\text{eff}}=\lambda_f+\lambda_b$ (effective decay constant)
λ_{eff}=ln2/10 d^{-1}
T=60 d

The effective dose during the first 60 d is given by

$$E_{60} = SN(60) = \frac{20 \cdot 10^{-3} \cdot \lambda_{\text{eff}} \cdot ALI}{ALI \cdot \lambda_{\text{eff}} \cdot 365}(60 + \frac{e^{-(\ln 2/10)60}}{(\ln 2/10)} - \frac{10}{\ln 2}) = 2.51 \cdot 10^{-3} \text{ Sv}$$

Here I_0 is set equal to ALI/365.

The effective dose for the time after 60 days is given by

$$E_\infty = SN_\infty = \frac{20 \cdot 10^{-3} \cdot ALI \cdot 10}{ALI \cdot 365 \cdot \ln 2}(1 - e^{-\frac{60 \cdot \ln 2}{10}}) = 0.78 \cdot 10^{-3} \, \text{Sv}$$

Answer. The effective dose during the first two months is 2.5 mSv and the total absorbed dose is 3.3 mSv.

Solution exercise 6.6.
The change of activity in the animal is given by the equation

$$\frac{\mathrm{d}A}{\mathrm{d}t} = I - \lambda_{\text{eff}} A \tag{6.3.18}$$

Solving the equation gives

$$A(t) = \frac{I}{\lambda_{\text{eff}}}(1 - e^{-\lambda_{\text{eff}}t}) \tag{6.3.19}$$

The total number of disintegrations in the animal is given by the sum of the number of disintegrations during the time T when the cattle is drinking the activated water and the number of disintegrations from the day they start to drink uncontaminated water to infinity, as the effective half life is short.

$$\tilde{A} = \int_0^T A(t)\,\mathrm{d}t + \int_0^\infty A(T)e^{-\lambda_{\text{eff}}t}\,\mathrm{d}t \tag{6.3.20}$$

or

$$\tilde{A} = \int_0^T \frac{I}{\lambda_{\text{eff}}}(1 - e^{-\lambda_{\text{eff}}t})\,\mathrm{d}t + \int_0^\infty \frac{I}{\lambda_{\text{eff}}}(1 - e^{-\lambda_{\text{eff}}T})e^{-\lambda_{\text{eff}}t}\,\mathrm{d}t \tag{6.3.21}$$

Solving the integral gives

$$\tilde{A} = \frac{I}{\lambda_{\text{eff}}}(T + \frac{e^{-\lambda_{\text{eff}}T}}{\lambda_{\text{eff}}} - \frac{1}{\lambda_{\text{eff}}}) + \frac{I}{\lambda_{\text{eff}}^2}(1 - e^{-\lambda_{\text{eff}}T}) \tag{6.3.22}$$

This can be simplified to

$$\tilde{A} = (I \cdot T)/\lambda_{\text{eff}} \tag{6.3.23}$$

Data:
$T=40$ d (time during which the animals are drinking contaminated water)
$\lambda_f = \ln 2/(12.3 \cdot 365.25) = 1.54 \cdot 10^{-4}$ d^{-1} (physical half life of ^{3}H)
$\lambda_b = 70/260$ d$^{-1} = 0.2693$ d^{-1} (biological half life of hydrogen)

$\lambda_{\text{eff}}=0.2694\,\text{d}^{-1}$ (effective half life of ^{3}H)
$I=70{\cdot}390\,\text{kBq}\,\text{d}^{-1}$ (intake per day if physical decay is neglected)

Data inserted in Eq. 6.3.23 gives

$$\tilde{A} = \frac{70\cdot390\cdot10^3\cdot40\cdot3600\cdot24}{0.2694} = 3.502\cdot10^{14}\,\text{Bqs}$$

^{3}H decays emitting low energetic β-radiation. All energy can then be assumed to be absorbed in the animal.

The absorbed dose is then given by

$$D = \frac{\tilde{A}_h\bar{E}}{m} \tag{6.3.24}$$

Data:
$\bar{E}=0.00568\,\text{MeV}$ (mean energy of the β-radiation)
$m=400\,\text{kg}$ (mass of the cattle)

Data inserted in Eq. (6.3.24) gives

$$D = \frac{3.502\cdot10^{14}\cdot0.00568\cdot1.602\cdot10^{-13}}{400} = 7.97\cdot10^{-4}\,\text{Gy}$$

If the physical decay of ^{3}H is not included, the absorbed dose will increase with less than 0.1 %.

Answer: The absorbed dose to the animal is 0.80 mGy.

Solution exercise 6.7.
The radiological limit for a radionuclide is given by ALI. From the definition of ALI the following relation is obtained

$$E = \int_0^{50} S\cdot ALIe^{-\lambda_{\text{eff}}t}\,dt = \frac{S\cdot ALI}{\lambda_{\text{eff}}}(1 - e^{-\lambda_{\text{eff}}\cdot50}) \tag{6.3.25}$$

S in the equation above is the relation effective dose to cumulated activity (Sv Bq^{-1} s^{-1}), which in our case is the same as the absorbed dose per decay.

As ^{63}Ni is emitting only β-radiation all energy may be assumed to be absorbed in the body and then S is obtained as $S = \bar{E}_\beta/m$ if the radiation weighting factor w_R

and the tissue weighting factor w_T are assumed to be equal to unity. $\bar{E}_\beta$ is the mean energy of the β-radiation, and m is the mass of the reference man.

Thus the equation may be expressed as

$$E = \frac{\bar{E}_\beta ALI}{m\lambda_{eff}}(1 - e^{-\lambda_{eff}\cdot 50}) \tag{6.3.26}$$

Data:
$\bar{E}_\beta$=17.1 keV (mean energy of the β-particles)
m=70 kg (mass of the reference man)
$\lambda_b = \ln 2/2.0\,\mathrm{y}^{-1}$ (biological decay constant)
$\lambda_f = \ln 2/96\,\mathrm{y}^{-1}$ (physical decay constant)
$\lambda_{eff} = \ln 2/2.0 + \ln 2/96 = 0.3538\,\mathrm{y}^{-1}$ (effective decay constant)
E=20.0 mSv (permissible effective dose per year for radiological personal over several years)

Data inserted in Eq. (6.3.26) gives

$$20 \cdot 10^{-3} = \frac{17.1 \cdot 1.602 \cdot 10^{-16} \cdot ALI \cdot (3600 \cdot 24 \cdot 365.25)}{70 \cdot 0.3538}(1 - e^{-0.3538 \cdot 50})$$

ALI=5.73·10⁶ Bq

The limit for chemical uptake of air with a concentration of $C\,\mathrm{kg\,m^{-3}}$ gives the yearly intake of nickel carbonyl

$$m_{Ni} = v\rho_{air}Cf \tag{6.3.27}$$

where
$v = 10.0 \cdot 10 \cdot 365\,\mathrm{m}^3$ (inhaled air volume per year)
$\rho_{air} = 1.20 \cdot 10^{-3}\,\mathrm{kg\,dm^{-3}}$ (t=20 °C, p=101.3 kPa)
$C = 1.0 \cdot 10^{-12}$ (concentration of Ni in air per mass unit)
f=0.5 (fraction of Ni absorbed in the body)
This gives

$m_{Ni} = 1.0 \cdot 10^4 \cdot 365 \cdot 1.20 \cdot 10^{-3} \cdot 1.0 \cdot 10^{-12} \cdot 0.5 = 2.19 \cdot 10^{-9}\,\mathrm{kg}$ (mass of nickel carbonyl per year)

The total molecule mass of the nickel carbonyl ($Ni(CO)_4$)) molecule is (63+4(12+16))=175

The number of nickel atoms is obtained by the relation using Avogadro's number.

$$N_{Ni} = \frac{m_{Ni}N_A}{m_a}$$

Data inserted gives

$$N_{Ni} = \frac{2.19 \cdot 10^{-9} \cdot 6.023 \cdot 10^{26}}{175}$$

The activity is then obtained by using the relation $A = \lambda \cdot N$

$$A_{Ni} = \frac{2.19 \cdot 10^{-9} \cdot 6.023 \cdot 10^{26}}{175} \frac{\ln 2}{96 \cdot 365.25 \cdot 24 \cdot 3600}$$

$A_{Ni} = 1.72 \cdot 10^6$ Bq

This activity is smaller than ALI. Thus the chemical toxicity sets the limit.

Answer: ALI is 5.73 MBq and the chemical limit corresponds to an activity of 1.72 MBq, and thus the experiment is acceptable from radiological point of view.

Solution exercise 6.8.

a) Single intake

The total number N_a of decays during time T after an intake A_0 is given by

$$N_a = \int_0^T A_0 e^{-\lambda_{eff} t}\, dt = \frac{A_0}{\lambda_{eff}}(1 - e^{-\lambda_{eff} T}) \qquad (6.3.28)$$

b) Continuous intake

The activity in the body at time t after a continuous intake I is given by

$$A = \frac{I}{\lambda_{eff}}(1 - e^{-\lambda_{eff} t}) \qquad (6.3.29)$$

The total number of decays N_b during time T is given by

$$N_b = \int_0^T \frac{I}{\lambda_{eff}}(1 - e^{-\lambda_{eff} t})\, dt \qquad (6.3.30)$$

$$N_b = \frac{I}{\lambda_{eff}}(T + \frac{e^{-\lambda_{eff} T}}{\lambda_{eff}} - \frac{1}{\lambda_{eff}}) \qquad (6.3.31)$$

Data:
$A_0 = 0.60 \cdot 13000$ Bq (0.60 kg with a concentration of 13000 Bq kg^{-1})

T=365 d (integration time)

$\lambda_f = \frac{\ln 2}{100}$ d^{-1} (physical decay constant)

$\lambda_b = \frac{\ln 2}{30 \cdot 365}$ d^{-1} (biological decay constant)

$\lambda_{eff} = \lambda_b + \lambda_f = 6.995 \cdot 10^{-3}$ d^{-1}=8.096·10^{-8} s^{-1} (effective decay constant)

I=50 Bq d^{-1} (continuous intake of activity; 0.5 dm^3 per day with a concentration of 100 Bq dm^{-3})

E=0.946·0.1734+0.054·0.4346+0.662·0.946=0.813 MeV (energy emitted per decay)

Data inserted in Eq.(6.3.28) and (6.3.31) and multiplying with the energy per decay gives

a) Single intake

$$E_a = \frac{0.813 \cdot 1.602 \cdot 10^{-13} \cdot 0.60 \cdot 13000}{8.096 \cdot 10^{-8}}(1 - e^{-6.995 \cdot 10^{-3} \cdot 365}) = 1.16 \cdot 10^{-2} \text{ J}$$

b) Continuous intake

$$E_b = \frac{0.813 \cdot 1.602 \cdot 10^{-13} \cdot 50}{8.096 \cdot 10^{-8} \cdot 3600 \cdot 24}(365 \cdot 24 \cdot 3600 + \frac{e^{-6.995 \cdot 10^{-3} \cdot 365}}{8.096 \cdot 10^{-8}} - \frac{1}{8.096 \cdot 10^{-8}})$$

$E_b = 1.88 \cdot 10^{-2}$ J

Answer: The emitted energy the first year for the single intake is 12 mJ and for the continuous intake 19 mJ.

Solution exercise 6.9.

The retention equation

$$R(t) = 0.10 \cdot e^{-0.347t} + 0.90 \cdot e^{-0.00630t} \quad (t \text{ in days}) \qquad (6.3.32)$$

can be considered as if the activity of Cs is taken up into two compartments with the fractions f_1 and f_2, with the respective biological decay constants $\lambda_{b,1}$ and $\lambda_{b,2}$.

For each compartment the activity at time t is then obtained from

$$\frac{dA}{dt} = fI - \lambda_{eff}A \qquad (6.3.33)$$

I is the activity intake per time unit.

Solving the equation gives

$$A = \frac{fI}{\lambda_{eff}}(1 - e^{\lambda_{eff}t}) \qquad (6.3.34)$$

The activity after 80 years should be less than 30 000 Bq. As the biological half life is short (<100 d), equilibrium can be assumed and the total activity is obtained from

$$A_{tot} = \frac{f_1 I}{\lambda_{eff,1}} + \frac{f_2 I}{\lambda_{eff,2}} \tag{6.3.35}$$

Data:
$f_1 = 0.10$, $\lambda_{b,1} = 0.347\,d^{-1}$
$f_2 = 0.90$, $\lambda_{b,2} = 0.00630\,d^{-1}$
$\lambda_f = \ln 2/(365.25 \cdot 30)\,d^{-1}$
$\lambda_{eff,1} = 0.347 + \ln2/(365.25 \cdot 30) = 0.347\,d^{-1}$
$\lambda_{eff,2} = 0.00630 + \ln2/(365.25 \cdot 30) = 0.00636\,d^{-1}$
$A_{tot} = 30\,000\,Bq$

Data inserted in Eq. 6.3.35 gives

$$30000 = \frac{0.1 \cdot I}{0.347} + \frac{0.9 \cdot I}{0.00636} \tag{6.3.36}$$

$I = 211.6\,Bq\,d^{-1}$

The concentration of the activity in the food is $300\,Bq\,kg^{-1}$. This gives a maximum intake of food per day of m=211.6/300=0.71 kg.

Answer: The maximum intake of the contaminated food should be $0.71\,kg\,d^{-1}$.

Solution exercise 6.10.
The activity in the water at time t is

$$A_w = A_{w,0} e^{-\lambda_{eff,w} t} \tag{6.3.37}$$

The activity in the cow at time t is obtained from

$$\frac{dA_c}{dt} = I - \lambda_{eff,c} A_c \tag{6.3.38}$$

I is the intake of activity by the cow per day. I is varying with time as the water activity decreases. Thus

$$I = I_0 e^{-\lambda_{eff,w} t} \tag{6.3.39}$$

Solution of Eq. (6.3.38) and (6.3.39) gives the variation with time of the activity in the cow.

$$A_c = \frac{I_0}{\lambda_{eff,c} - \lambda_{eff,w}} (e^{-\lambda_{eff,w} t} - e^{-\lambda_{eff,c} t}) \tag{6.3.40}$$

Maximal activity is obtained by derivation of Eq. (6.3.40)

$$\frac{dA_c}{dt} = \frac{I_0}{\lambda_{eff,c} - \lambda_{eff,w}}(-\lambda_{eff,w}e^{-\lambda_{eff,w}t} + \lambda_{eff,c}e^{-\lambda_{eff,c}t}) \tag{6.3.41}$$

Maximum is obtained when $\frac{dA_c}{dt} = 0$. Thus

$$\lambda_{eff,w}e^{-\lambda_{eff,w}t} = \lambda_{eff,c}e^{-\lambda_{eff,c}t} \tag{6.3.42}$$

and

$$t_{max} = \frac{\ln\frac{\lambda_{eff,c}}{\lambda_{eff,w}}}{\lambda_{eff,c} - \lambda_{eff,w}} \tag{6.3.43}$$

Thus $A_{c,max}$ is obtained from

$$A_{c,max} = \frac{I_0}{\lambda_{eff,c} - \lambda_{eff,w}}(e^{-\lambda_{eff,w}t_{max}} - e^{-\lambda_{eff,c}t_{max}}) \tag{6.3.44}$$

Data:
$I_0 = 70 \cdot 2.5 \cdot 10^3 \, \text{Bq} \, \text{d}^{-1}$ (initial activity intake in cow per day)

$\lambda_{eff,w} = \frac{1000}{100000} + \frac{\ln 2}{30 \cdot 365.25} = 0.01006 \, \text{d}^{-1}$ (effective decay constant in water)

$\lambda_{eff,c} = \frac{\ln 2}{100} + \frac{\ln 2}{30 \cdot 365.25} = 6.995 \cdot 10^{-3} \, \text{d}^{-1}$ (effective decay constant in cow)

This inserted in Eq. 6.3.43 gives

$$t_{max} = \frac{\ln\frac{6.995 \cdot 10^{-3}}{0.01006}}{6.995 \cdot 10^{-3} - 0.01006} = 118.56 \, \text{d} \tag{6.3.45}$$

Data inserted in Eq. (6.3.44) for $A_{c,max}$ gives

$$A_{c,max} = \frac{70 \cdot 2.5 \cdot 10^3}{6.995 \cdot 10^{-3} - 0.01006}(e^{-0.01006 \cdot 118.56} - e^{-6.995 \cdot 10^{-3} \cdot 118.56})$$

$A_{c,max} = 7.59 \cdot 10^6 \, \text{Bq}$

Answer: The maximal activity in the cow is 7.6 MBq.

Solution exercise 6.11.
The activity in the body at time t is given by

$$A = \frac{I}{\lambda_{eff}}(1 - e^{-\lambda_{eff}t}) \tag{6.3.46}$$

where I is the intake per time unit and λ_{eff} is the effective decay constant.

After some months, equilibrium is obtained and then the equation is reduced to

$$A = \frac{I}{\lambda_{\text{eff}}} \tag{6.3.47}$$

^{3}H disintegrates by emitting low energetic β-radiation. Thus all energy may be assumed to be absorbed in the body.

The absorbed dose during a time t is then obtained from the relation

$$D = \frac{A\bar{E}_\beta}{m}t \tag{6.3.48}$$

Data:
I=2·800 Bq d^{-1} (intake in the body per day)
λ_{eff}=ln2/10 d^{-1} (physical half life may be neglected (12.3 y)).
$\bar{E}_\beta$=5.68·10^{-3} · 1.602 · 10^{-13} J (mean energy per decay)
m=70 kg (body mass)

Data inserted in Eq. (6.3.48) gives the dose during a year when equilibrium has been obtained

$$D = \frac{2 \cdot 800 \cdot 10 \cdot 5.68 \cdot 10^{-3} \cdot 1.602 \cdot 10^{-13} \cdot 365 \cdot 24 \cdot 3600}{\ln 2 \cdot 70} = 9.46 \cdot 10^{-6} \text{ Gy}$$

During the first year a lower dose is obtained as it takes some time to obtain the equilibrium activity in the body and the cumulated activity is then obtained according to the relation

$$\tilde{A} = \int_0^{365} \frac{I_0}{\lambda_{\text{eff}}}(1 - e^{-\lambda_{\text{eff}}t})dt = \frac{I_0}{\lambda_{\text{eff}}}[t + \frac{e^{-\lambda_{\text{eff}}t}}{\lambda_{\text{eff}}}]_0^{365} \tag{6.3.49}$$

Data inserted gives

$$\tilde{A} = \frac{2 \cdot 800 \cdot 10}{\ln 2}[365 \cdot 24 \cdot 3600 + \frac{e^{-365 \cdot \ln 2/10}}{\ln 2/(10 \cdot 24 \cdot 3600)} - \frac{1}{\ln 2/(10 \cdot 24 \cdot 3600)}]$$

$\tilde{A}$ = 6.992 · 10^{11} Bq s

The absorbed dose is obtained from the relation

$$D = \frac{\tilde{A}\bar{E}_\beta}{m} \tag{6.3.50}$$

Inserted data gives

D=9.09·10^{-6} Gy

The effective dose is obtained by multiplying with the radiation quality factor w_R and the tissue weighting factor w_T. As ^{3}H is decaying by emitting β-radiation and the dose distribution is homogeneous, both factors are equal to unity and the effective dose is equal to absorbed dose.

Answer: The effective dose during the first year is 9.1 μSv and in the future 9.5 μSv per year.

Solution exercise 6.12.
Assuming continuous release of activity, the activity concentration in the room is given by

$$C(t) = \frac{I}{V\lambda_{\text{room}}}(1 - e^{-\lambda_{\text{room}}t}) \tag{6.3.51}$$

where I is the release rate of activity to the air, V is the volume of the room and λ_{room} is the rate of change in the activity in the room due to ventilation and decay.

The activity, A_{Th}, in the thyroid is then given by

$$\frac{dA_{\text{Th}}}{dt} = U_{\text{Th}} - \lambda_{\text{eff}}A_{\text{Th}} \tag{6.3.52}$$

U_{Th} is the uptake in the thyroid per time unit and is given by

$$U_{\text{Th}} = \frac{Ihf}{V\lambda_{\text{room}}}(1 - e^{-\lambda_{\text{room}}t}) \tag{6.3.53}$$

where h is the breathing rate, f is part of activity taken up in the thyroid, λ_{eff} is the effective decay constant in the thyroid, and λ_{room} is the effective ventilation rate constant in the room.

Inserting U_{Th} in the differential equation (6.3.52), rearranging and integrating gives

$$\int\left(\frac{dA_{\text{Th}}}{dt} + \lambda_{\text{eff}}A_{\text{Th}}\right)dt = \int\frac{Ihf}{V\lambda_{\text{room}}}(1 - e^{-\lambda_{\text{room}}t})\,dt \tag{6.3.54}$$

Multiplying both sides with the integrating factor $e^{\lambda_{\text{eff}}t}$ and solving the integrals give

$$A_{\text{Th}} \cdot e^{\lambda_{\text{eff}}t} = \frac{Ihf}{V\lambda_{\text{room}}}\left(\frac{e^{\lambda_{\text{eff}}t}}{\lambda_{\text{eff}}} - \frac{e^{(\lambda_{\text{eff}}-\lambda_{\text{room}})t}}{\lambda_{\text{eff}} - \lambda_{\text{room}}}\right) + C \tag{6.3.55}$$

The constant C is obtained by assuming that the activity $A_{\text{Th}}=0$, when $t=0$.

$$C = \frac{Ihf}{V\lambda_{\text{room}}}\left(-\frac{1}{\lambda_{\text{eff}}} + \frac{1}{\lambda_{\text{eff}} - \lambda_{\text{room}}}\right) \tag{6.3.56}$$

Inserting C in the equation and rearranging gives

$$A_{Th} = \frac{Ihf}{V\lambda_{room}}\left(\frac{1 - e^{-\lambda_{eff}t}}{\lambda_{eff}} + \frac{e^{-\lambda_{eff}t} - e^{-\lambda_{room}t}}{\lambda_{eff} - \lambda_{room}}\right) \qquad (6.3.57)$$

Data:

I=0.20 MBq h^{-1} (release of activity to air)

V=20 m^3 (volume of the room)

h=0.020·60 m^3 h^{-1} (breathing rate)

f=0.30 (uptake in the thyroid)

t=8.0 h (working time)

T_b=75 d (biological half life)

T_f=8.04 d (physical half life)

λ_v=0.50 h^{-1} (room ventilation rate)

$\lambda_{room} = 0.50 + \frac{\ln 2}{8.04\cdot 24} = 0.504 \, h^{-1}$

$\lambda_{eff} = \frac{\ln 2}{75\cdot 24} + \frac{\ln 2}{8.04\cdot 24} = 3.98\cdot 10^{-3} \, h^{-1}$

Data inserted in Eq. 6.3.57 gives

$$A_b = \frac{0.20 \cdot 10^6 \cdot 0.020 \cdot 60 \cdot 0.30}{20 \cdot 0.504}\left(\frac{1 - e^{-3.98\cdot 10^{-3}\cdot 8}}{3.98 \cdot 10^{-3}} + \frac{e^{-3.98\cdot 10^{-3}\cdot 8} - e^{-0.504\cdot 8}}{3.98 \cdot 10^{-3} - 0.504}\right)$$

A_b=42.7 kBq

Answer: The activity in the thyroid is 43 kBq.

6.3.2 Point Radioactive Sources

Solution exercise 6.13.

The absorbed dose rate at the calculation point P is, assuming charged particle equilibrium, given by the relation

$$\dot{D} = \frac{Af}{4\pi r^2}B(\mu_{en}/\rho)_{water}h\nu e^{-\mu_{Fe}d_{Fe}}e^{-\mu_{concrete}d_{concrete}} \qquad (6.3.58)$$

Data:

A=0.63 GBq (source activity)

B=3.0 (dose build-up factor)

f=2.0 (number of photons per decay)

$h\nu$=1.25 MeV (mean photon energy for ^{60}Co)

r=60 cm (distance source-calculation point)

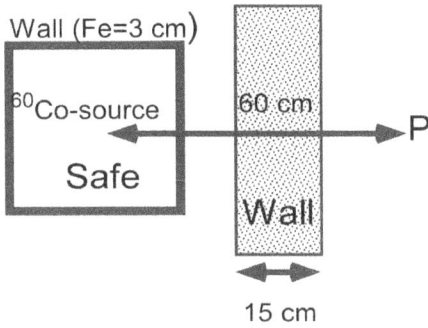

Figure 6.10: Illustration of the irradiation geometry in exercise 6.13.

d_{Fe}=3.0 cm (safe wall thickness)
$d_{concrete}$=15.0 cm (concrete wall thickness)
$(\mu/\rho)_{Fe}$=0.00535 m^2 kg^{-1} (mass attenuation coefficient for iron)
$\rho_{Fe} = 7.86 \cdot 10^3$ kg m^{-3} (density of iron)
$(\mu/\rho)_{concrete}$=0.005807 m^2 kg^{-1} (mass attenuation coefficient for concrete)
$\rho_{concrete} = 2.35 \cdot 10^3$ kg m^{-3} (density of concrete)
$(\mu_{en}/\rho)_{water}$=0.00296 m^2 kg^{-1} (mass energy absorption coefficient for water)

Data inserted in Eq. 6.3.58 gives

$$\dot{D} = \frac{0.63 \cdot 10^9 \cdot 2 \cdot 3.0 \cdot 0.00296 \cdot 1.25 \cdot 1.602 \cdot 10^{-13}}{4\pi \cdot 0.60^2}$$
$$\times e^{-0.00535 \cdot 7.86 \cdot 10^3 \cdot 0.03} \cdot e^{-0.005807 \cdot 2.35 \cdot 10^3 \cdot 0.15} = 1.81 \cdot 10^{-8} \text{ Gy s}^{-1}$$

During a week the absorbed dose will be

D=1.8·10^{-8} · 40 · 3600= 2.6·10^{-3} Gy

Answer: The absorbed dose during one week will be 2.6 mGy. This is not acceptable according to the ICRP recommendations.

Solution exercise 6.14.

The absorbed dose rate at point P outside the safe (see Fig. 6.11) is, assuming charged particle equilibrium, given by the equation

$$\dot{D} = \frac{Af}{4\pi r^2} B(\mu d)(\mu_{en}/\rho)_{water} h \nu e^{-\mu_{Fe} d_{Fe}} \qquad (6.3.59)$$

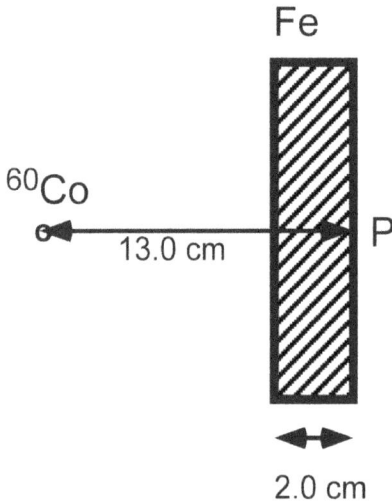

Figure 6.11: Illustration of the irradiation geometry in exercise 6.14.

$B(\mu d)$ is the build-up factor given by the Berger expression

$$B = 1 + a\mu d e^{b\mu d} \tag{6.3.60}$$

Data:
$\dot{D}$=10 μGy h^{-1} (dose rate at point P)
$h\nu$=1.25 MeV (mean photon energy for ^{60}Co)
f=2.0 (number of photons per decay)
r=13.0 cm (distance between source and calculation point)
d_{Fe}=2.0 cm (iron thickness)
$(\mu/\rho)_{Fe}$ = 0.00535 m^2 kg^{-1} (mass attenuation coefficient for Fe)
ρ_{Fe} = 7.86 · 10^3 kg m^{-3} (density of Fe)
$(\mu_{en}/\rho)_{water}$=0.00296 m^2 kg^{-1} (mass energy absorption coefficient for water)
a=0.955 and b=0.024 (parameters for Berger expression)

Data inserted in Eq. (6.3.59) and (6.3.60)

$$\frac{10 \cdot 10^{-6}}{3600} = \frac{A \cdot 2.0}{4\pi(0.130)^2}(1 + 0.955 \cdot 0.020 \cdot 0.00535 \cdot 7.86 \cdot 10^3 \cdot e^{0.024 \cdot 7.86 \cdot 10^3 \cdot 0.00535 \cdot 0.020}) \times$$
$$e^{-7.86 \cdot 10^3 \cdot 0.00535 \cdot 0.020} \cdot 0.00296 \cdot 1.25 \cdot 1.602 \cdot 10^{-13}$$

Solving the equation gives

A=0.63 MBq

Answer: The maximum activity of ^{60}Co is 0.63 MBq.

Solution exercise 6.15.

Method 1:

The kerma in air at P (see Fig. 6.12) is given by the equation

$$K_1 = \frac{A \sum f_i h v_i (\mu_{tr}/\rho)_{air,i}}{4\pi r_1^2} B(\mu d) t_1 e^{-\mu_{Pb,i} d} \tag{6.3.61}$$

B is the build-up factor given by the Berger expression

$$B = 1 + a\mu d e^{b\mu d} \tag{6.3.62}$$

Method 2:

The kerma in air at P is now given by the equation

$$K_2 = \frac{A \sum f_i h v_i (\mu_{tr}/\rho)_{air,i}}{4\pi r_2^2} t_2 \tag{6.3.63}$$

The ratio of the kerma values is given by

$$\frac{K_1}{K_2} = \frac{\sum f_i h v (\mu_{tr}/\rho)_{air,i} 4\pi r_2^2 B(\mu d) t_1 e^{-\mu_{Pb,i} d_{Pb}}}{A \sum f_i h v (\mu_{tr}/\rho)_{air,i} 4\pi r_1^2 t_2} \tag{6.3.64}$$

With only one energy the relation can be simplified to

$$\frac{K_1}{K_2} = \frac{r_2^2 B(\mu d) t_1 e^{-\mu_{Pb} d_{Pb}}}{r_1^2 t_2} \tag{6.3.65}$$

Data:
A=200 GBq (activity of the ^{60}Co-source)
hv=1.25 MeV (mean photon energy for ^{60}Co)
f=2.0 (number of photons per decay)
r_1=50 cm, r_2=40 cm (distances between source and measuring point)
t_1=600 s, t_2=140 s (time for measurement)
d_{Pb}=4.0 cm (thickness of lead absorber)
$(\mu/\rho)_{Pb}$=0.00588 m^2 kg^{-1} (mass attenuation coefficient for lead)
ρ_{Pb} = 11.35 · 10^3 kg m^{-3} (density of lead)
$(\mu_{tr}/\rho)_{air}$=0.00267 m^2 kg^{-1} (mass energy transfer coefficient for air)
a=0.33 and b=-0.011 (parameters for Berger expression)

Data inserted in Eq. 6.3.62 gives

$$B = 1 + 0.33 \cdot 0.00588 \cdot 0.040 \cdot 11.35 \cdot 10^3 \cdot e^{-0.011 \cdot 0.00588 \cdot 0.040 \cdot 11.35 \cdot 10^3} = 1.86$$

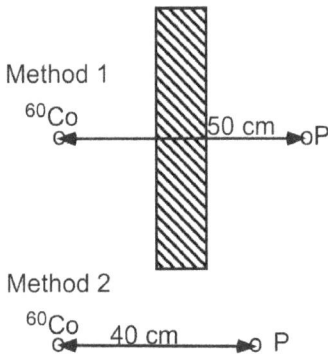

Figure 6.12: Illustration of the irradiation geometry in exercise 6.15.

Data inserted in Eq. 6.3.65 gives

$$\frac{K_1}{K_2} = \frac{40^2 \cdot 1.86 \cdot 600 \cdot e^{-0.00588 \cdot 0.040 \cdot 11.35 \cdot 10^3}}{50^2 \cdot 140} = 0.35$$

Method 1 gives the lowest kerma.

The kerma rate in air for method 1 is

$$\dot{K}_1 = \frac{200 \cdot 10^9 \cdot 2 \cdot 1.25 \cdot 1.602 \cdot 10^{-13} \cdot 0.00267 \cdot 1.86 \cdot e^{-0.00588 \cdot 0.040 \cdot 11.35 \cdot 10^3}}{4\pi \cdot 0.5^2}$$

$\dot{K}_1 = 8.752 \cdot 10^{-6}\,\mathrm{Gy\,s^{-1}} = 32\,\mathrm{m\,Gy\,h^{-1}}$

Answer: Method 1 gives lower kerma. The kerma rate in air for this method is $32\,\mathrm{m\,Gy\,h^{-1}}$.

Solution exercise 6.16.
The air kerma at the father is obtained both from scattered radiation and from leakage radiation.

I. Scattered radiation

The air kerma for scattered radiation is given by

$$K_S = \frac{\dot{K}tS}{d^2} \tag{6.3.66}$$

where
$t = 5 \cdot 4 \cdot 150 = 3000\,\mathrm{s}$ (total treatment time)
$S = 0.001$ (fraction of scattered radiation 1.0 m from the center of the scatterer)

d=4.0 m (distance from scatterer to the father)

$\dot{K}$=air kerma rate at the center of the patient given by

$$\dot{K} = \frac{21.8 \cdot 10^{-3} \cdot 1.0^2}{0.9^2} \; \text{Gy s}^{-1}$$

Data inserted gives for the scattered radiation

$$K_S = \frac{21.8 \cdot 10^{-3} \cdot 1.0^2 \cdot 3000 \cdot 0.001}{0.9^2 \cdot 4.0^2} \; \text{Gy}$$

The relation, effective dose to air kerma, is 1.09 (given in the exercise text). Thus, the effective dose E_S to the father from the scattered radiation is

$$E_S = K_S \cdot 1.09 \tag{6.3.67}$$

The permissible effective dose E_P to the father shall be below 100 μSv.

The necessary reduction of the kerma is then given by the transmission factor

$$B = \frac{E_P}{E_S} \tag{6.3.68}$$

With inserted data

$$B = \frac{0.9^2 \cdot 4^2 \cdot 100 \cdot 10^{-6}}{21.8 \cdot 10^{-3} \cdot 1.0^2 \cdot 3000 \cdot 0.001 \cdot 1.09} = 1.82 \cdot 10^{-2}$$

From transmission data (Fig 26, ICRP 33 (ICRP, 1982)) a thickness of 1.4 cm Pb is obtained.

II. Leakage radiation.

According to ICRP 33 the air kerma rate 1.0 m from the source should be below 0.1% of the primary air kerma rate at 1.0 m.

$$\dot{K}_L = 21.8 \cdot 10^{-3} \cdot 0.1 \cdot 10^{-2}$$

The effective dose to the father from leakage radiation without shielding during the total treatment is then given by

$$E_L = \frac{21.8 \cdot 10^{-3} \cdot 0.1 \cdot 10^{-2} \cdot 3000 \cdot 1.09}{4^2} = 4.46 \, \text{mSv}$$

To calculate the necessary absorber thickness, the number of TVT (tenth value thicknesses) is calculated.

The number of TVT is obtained by

$$N_{TVT} = ^{10}\!\log \frac{4.46 \cdot 10^{-3}}{100 \cdot 10^{-6}} = 1.65$$

According to Table 4 (ICRP 33) 1 TVT=4.0 cm Pb.

This gives the necessary thickness of lead for the leakage radiation to 4.0·1.65=6.60 cm.

The necessary thickness for the scattered radiation is 1.4 cm Pb. This is more than 1 TVT smaller than for leakage radiation. Thus 6.60 cm is enough according the recommendations of ICRP.

Answer: The necessary thickness of lead is 6.6 cm.

Solution exercise 6.17.
Wall at point A:

The wall is hit by primary radiation and the transmission is obtained by the relation

$$B = \frac{Pd_p^2}{W} \tag{6.3.69}$$

where
$P = 10 \cdot 10^{-6} \, \text{Gy} \, \text{h}^{-1}$ (permissible absorbed dose rate taking use and occupancy factors into account)
d_p=5.5 m (distance target-point A)
W=4.0 Gy min^{-1} (primary absorbed dose rate)

Data inserted in Eq. 6.3.69 gives

$$B = \frac{10 \cdot 10^{-6} \cdot 5.5^2}{60 \cdot 4.0} = 1.26 \cdot 10^{-6} \tag{6.3.70}$$

From Fig. 13 in ICRP 33 (ICRP, 1982) the necessary wall thickness is obtained and around 190 cm concrete.

Wall at point B:

This wall is not hit by primary radiation but only scattered and leakage radiation. The transmission for scattered radiation is obtained from the relation

$$B = \frac{Pd_s^2}{WS} \tag{6.3.71}$$

where
$P = 100 \cdot 10^{-6} \, \text{Gy} \, \text{h}^{-1}$ (permissible absorbed dose rate taking use and occupancy factors into account)
d_s=5.0 m (distance isocenter-point B)
$S = 0.03 \cdot 10^{-2}$ (fraction of scattered radiation)

Data inserted in Eq. 6.3.71 gives

$$B = \frac{100 \cdot 10^{-6} \cdot 5.0^2}{60 \cdot 4.0 \cdot 0.03 \cdot 10^{-2}} = 3.47 \cdot 10^{-2} \qquad (6.3.72)$$

From Fig. 27 in ICRP 33 the necessary wall thickness is obtained and around 32 cm concrete.

The number of tenth value thicknesses for the leakage radiation is given by

$$N_{TVT} =^{10} \log \frac{W_L}{d_s^2 P} \qquad (6.3.73)$$

where $W_L = 0.1 \cdot 10^{-2} \cdot 4.0$ Gy min^{-1} (absorbed dose rate 1.0 m from isocenter). In principle the distance should be calculated from the target, but often the distance from isocenter to calculation point is used.
The other parameters are the same as above. Data inserted gives

$$N_{TVT} =^{10} \log \frac{0.001 \cdot 4.0}{5.0^2 \cdot 100 \cdot 10^{-6}/60} =^{10} \log 96 = 1.98 \qquad (6.3.74)$$

Table 3 in ICRP 33 gives that 1 TVT=33.8 cm concrete. Thus the needed wall thickness is 1.98x33.8=67 cm

The needed wall thickness for leakage radiation is more than 1 TVT larger than the wall thickness for scattered radiation. Thus 67 cm is enough.

Answer: The wall thickness at point A should be 190 cm and for point B 67 cm concrete.

Solution exercise 6.18.
The kerma in water at the patient is given by

$$K = \frac{A\Gamma}{r^2} B \cdot (1 - f) \cdot (\mu_{tr}/\rho)_{water,air} \cdot t \qquad (6.3.75)$$

t=50·3600 s (treatment time)
f=0.30 (absorption in the brachytherapy patient)
E=0.02 mSv (effective dose during one week)
K=0.02 mGy (kerma during one week as $w_R=w_T=1$)
Γ=20.3·10^{-18} Gy s^{-1} Bq^{-1} m^2 (air kerma rate constant)
$(\mu_{tr}/\rho)_{air}$=0.00294 m^2 kg^{-1} (mass energy transfer coefficient for air)
$(\mu_{tr}/\rho)_{water}$=0.00327 m^2 kg^{-1} (mass energy transfer coefficient for water)
B=transmission through the wall
A=2.60 GBq (activity of the ^{137}Cs source)

Data inserted gives

$$0.02 \cdot 10^{-3} = \frac{2.60 \cdot 10^9 \cdot 20.3 \cdot 10^{-18} \cdot 0.00327}{2^2 \cdot 0.00294} B \cdot (1 - 0.30) \cdot 50 \cdot 3600$$

$B=1.08 \cdot 10^{-2}$

Fig 18 in ICRP 33 (ICRP, 1982) gives 42 cm concrete.

Answer: The concrete wall thickness shall be 42 cm.

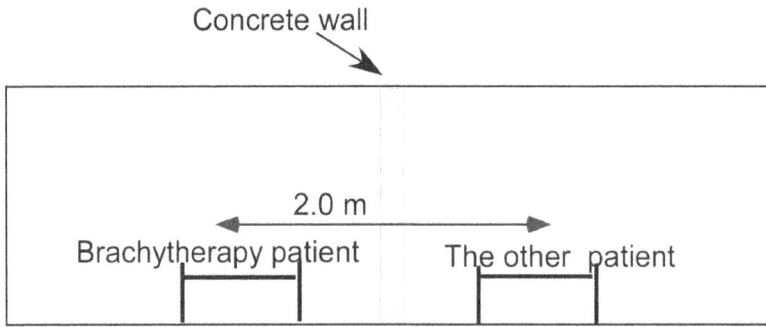

Figure 6.13: Illustration of the ward rooms with beds.

6.3.3 Extended Radioactive Sources

Solution exercise 6.19. The fluence rate $\dot{\Phi}$ from a line source (see Fig. 6.14) is given by

$$\dot{\Phi} = \frac{S_L f}{4\pi h}[F(|\theta_2|, \sum(\mu_i x_i)) + F(|\theta_1|, \sum(\mu_i x_i))] \tag{6.3.76}$$

where

$$S_L = \frac{A}{l}$$

is the line source strength and

$$F(|\theta|, \sum(\mu_i x_i))$$

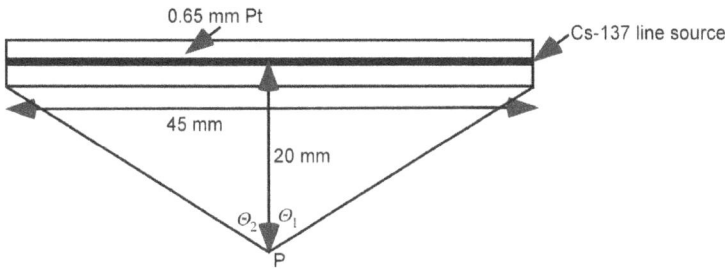

0.65 mm Pt

Cs-137 line source

45 mm

20 mm

θ_2 | θ_1

P

Figure 6.14: Brachytherapy line source in exercise 6.19.

is the Sievert integral.

If P is centrally located then $|\theta_2|=|\theta_1|$.

If charged particle equilibrium (CPE) is assumed, the absorbed dose rate $\dot{D}$ to water is given by

$$\dot{D} = \dot{\Phi}h\nu(\mu_{en}/\rho)_{water} \tag{6.3.77}$$

Data:
A=57.5 GBq (source activity)
l=45 mm (length of the brachytherapy source)
f=0.85 (number of 0.662 MeV photons per decay)
$h\nu$=0.662 MeV (photon energy)
h=20 mm (shortest distance to line source)
$\theta_2 = \theta_1$=0.844 rad (48.4°) (opening angle)
$(\mu_{en}/\rho)_{water}$=0.00325 m^2 kg^{-1} (mass energy absorption coefficient for water)
$\mu_{Pt} = (\mu/\rho)_{Pt} \cdot \rho_{Pt} = 0.01062 \cdot 21.45 \cdot 10^3$ m^{-1} (linear attenuation coefficient for Pt)
μ_{water}=8.60 m^{-1} (linear attenuation coefficient for water)
d_{water}=20-0.65=19.35 mm (water thickness)
d_{Pt}=0.65 mm (platina thickness)
$\sum(\mu_i d_i) = 8.60 \cdot 19.35 \cdot 10^{-3} + 0.01062 \cdot 21.45 \cdot 10^3 \cdot 0.65 \cdot 10^{-3}$=0.3145

F is obtained from Table 6.4 which gives F(0.844,0.3145)=0.593

Data inserted in Eq. (6.3.76) gives

$$\dot{\Phi} = \frac{57.5 \cdot 10^9 \cdot 0.85 \cdot 2 \cdot 0.593}{4\pi \cdot 0.045 \cdot 0.020} = 5.125 \cdot 10^{12} \text{ m}^{-2}\text{s}^{-1}$$

The absorbed dose rate is then

$$\dot{D} = 5.125 \cdot 10^{12} \cdot 0.662 \cdot 1.602 \cdot 10^{-13} \cdot 0.00325 = 1.766 \cdot 10^{-3} \text{ Gy s}^{-1}$$

The required absorbed dose is 2.0 Gy.

This gives the treatment time

$$t = \frac{2.0}{1.766 \cdot 10^{-3}} = 1133\,\text{s}$$

Answer: The treatment time is 18.9 min.

Solution exercise 6.20.
The fluence rate at the center of the ring is given by

$$\dot{\Phi} = \int\limits_0^{2\pi R} \frac{S_L f \, dx}{4\pi R^2} e^{-\mu d} = \frac{S_L f}{2R} e^{-\mu d} \tag{6.3.78}$$

The kerma rate to water is
$$\dot{K} = \dot{\Phi} h\nu (\mu_{tr}/\rho)_{\text{water}} \tag{6.3.79}$$

The total kerma during the time T is

$$K = \int\limits_0^T \dot{K} e^{-\lambda t} \, dt = \frac{\dot{K}}{\lambda}(1 - e^{-\lambda T}) \tag{6.3.80}$$

Combining Eq. (6.3.78), Eq. (6.3.79) and Eq. (6.3.80) gives

$$K = \frac{S_L f}{2R} e^{-\mu d} h\nu (\mu_{tr}/\rho)_{\text{water}} \frac{1}{\lambda}(1 - e^{-\lambda T}) \tag{6.3.81}$$

^{15}O decays by β^+-decay and the positrons are supposed to be annihilated, giving two annihilation photons in opposite directions.

Data:
$S_L = 1.30\,\text{GBq m}^{-1}$ (line source strength)
$f = 2.0$ (number of photons per annihilation)
$h\nu = 0.511\,\text{MeV}$ (energy of the annihilation photons)
$(\mu_{tr}/\rho)_{\text{water}} = 0.00330\,\text{m}^2\,\text{kg}^{-1}$ (mass energy transfer coefficient for water)
$(\mu/\rho)_{\text{Fe}} = 0.00834\,\text{m}^2\,\text{kg}^{-1}$ (mass attenuation coefficient for iron)
$\rho_{\text{Fe}} = 7.87 \cdot 10^3\,\text{kg m}^{-3}$ (density of iron)
$\lambda = \ln2/122.2\,\text{s}^{-1}$ (decay constant for ^{15}O)
$T = 30.0\,\text{min}$ (measurement time)
$R = 0.75\,\text{m}$ (betatron radius)
$d = 10\,\text{mm}$ (iron thickness)

Data inserted gives

$$K = \frac{1.30 \cdot 10^9 \cdot 2 \cdot 0.511 \cdot 1.602 \cdot 10^{-13} \cdot 0.00330 \cdot 122.2 \cdot e^{-0.00834 \cdot 0.010 \cdot 7.87 \cdot 10^3}}{2 \cdot 0.75 \cdot \ln 2}$$
$$\times (1 - e^{-\frac{30 \cdot 60 \cdot \ln 2}{122.2}}) \, \text{Gy}$$

$K = 4.28 \cdot 10^{-5}$ Gy

Answer: The kerma to a small mass of water at the center of the betatron is 0.043 mGy.

Solution exercise 6.21.
The air kerma rate at x (see Fig. 6.15) is given by

$$\dot{K} = \dot{\Phi} h\nu (\mu_{tr}/\rho)_{air} \qquad (6.3.82)$$

where

$$\dot{\Phi} = \frac{S_L f}{4\pi h}[F(|\theta_2|, \sum(\mu_i x_i)) + F(|\theta_1|, \sum(\mu_i x_i))] \qquad (6.3.83)$$

^{18}F decays by β^+-decay and the positrons are supposed to be annihilated, giving two

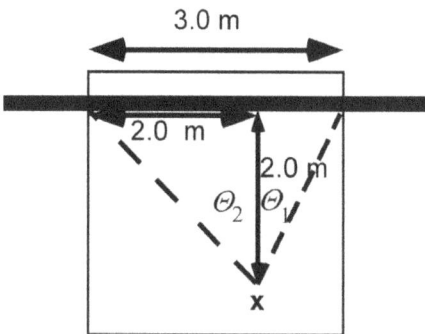

Figure 6.15: Sketch of the room with a tube containing a radioactive gas.

annihilation photons in opposite directions.

Data:
$S_L = 1.20 \cdot 10^9$ Bq m^{-1} (line source strength)
$f = 2.0$ (number of photons per annihilation)

hv=0.511 MeV (annihilation photon energy)

θ_1=26.56°=0.4635 rad, θ_2=45°=0.7854 rad (opening angles)

d_1=2.0 mm Fe, d_2=2.0 m air

μ_{Fe}=0.00834·7.86 · 10^3 m^{-1}

μ_{air}=0.00865·1.20 m^{-1}(assuming T=293 K and p=101.3 kPa)

$\sum(\mu_i d_i)$=0.131+0.021=0.152 (total attenuation thickness)

$(\mu_{tr}/\rho)_{air}$=0.00295 m^2 kg^{-1} (mass energy transfer coefficient for air)

F is obtained from Table 6.4 which gives

$F(45, 0.152)$=0.663

$F(26.56, 0.152)$=0.398

Data inserted in Eq. 6.3.82 gives

$$\dot{K} = \frac{1.20 \cdot 10^9 \cdot 2 \cdot 0.511 \cdot 1.602 \cdot 10^{-13} \cdot 0.00295}{4\pi \cdot 2.0}(0.663 + 0.398) \text{ Gy s}^{-1}$$

$$\dot{K} = 2.45 \cdot 10^{-8} \text{ Gy s}^{-1}=88 \,\mu\text{Gy h}^{-1}$$

Answer: The air kerma rate is 88 μGy h^{-1}. This kerma rate is too high.

Solution exercise 6.22.

The fluence rate decreases due to decay of the activity.

The kerma to tissue is thus given by integrating over time

$$K = \int_0^T \dot{\Phi}_0 hv(\mu_{tr}/\rho)_{tissue}e^{-\lambda t}\, dt \tag{6.3.84}$$

$$K = \dot{\Phi}_0 \cdot hv(\mu_{tr}/\rho)_{tissue}\frac{1}{\lambda}(1 - e^{-\lambda T}) \tag{6.3.85}$$

The initial fluence rate for an infinite surface area is given by

$$\dot{\Phi}_0 = \frac{S_A f}{2}E_1(\mu d) \tag{6.3.86}$$

where $E_1(\mu d)$ is the exponential integral of the first order.

Data:

S_A=450 MBq m^{-2} (area source strength)

f=0.812 (number of photons per decay)

T=14.0 d (integration time)

$h\nu$=0.364 MeV (main photon energy of ^{131}I decay)

$(\mu/\rho)_{air}$=0.00995 m^2 kg^{-1} (mass attenuation coefficient for air)

$\rho_{air} = (1.293 \cdot 100 \cdot 273)/(101.3 \cdot 288)$ kg/m^{-3}(density of air at T=288 K, p=100 kPa)

d=1.50 m (air thickness)

$\lambda_{I131} = \ln 2/(8.04 \cdot 24 \cdot 3600)$ s^{-1}

$(\mu_{tr}/\rho)_{tissue}$=0.00322 m^2 kg^{-1}

$(\mu d)_{air} = 0.0181$

$E_1(0.0181)$=3.462 (Table 6.6)

Data inserted in Eq. (6.3.85) gives

$$K = \frac{450 \cdot 10^6 \cdot 0.812 \cdot 3.462 \cdot 8.04 \cdot 24 \cdot 3600}{2 \cdot \ln 2} \cdot 0.00322 \cdot 0.364$$
$$\times 1.602 \cdot 10^{-13}(1 - e^{-\frac{14 \cdot \ln 2}{8.04}}) = 0.0834 \, \text{Gy}$$

Answer: The kerma to tissue is 83 mGy.

Solution exercise 6.23.

The fluence rate centrally above a circular radioactive area is given by the equation

$$\dot{\Phi} = \frac{S_A f}{2}[E_1(\sum \mu_i x_i) - E_1(\sum \mu_i x_i \sec \theta)] \tag{6.3.87}$$

where E_1 is the exponential integral of first order. β is the opening angle (see Fig. 6.16). The corresponding kerma rate in air is given by

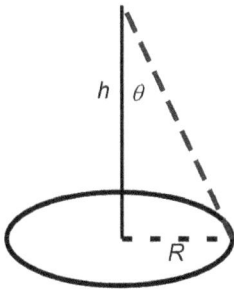

Figure 6.16: Circular surface area source.

$$\dot{K} = \dot{\Phi}\sum h\nu_i(\mu_{tr}/\rho)_{i,air} \tag{6.3.88}$$

Data:

S_A=80000 Bq m^{-2} (area source activity)

f=0.898·0.946 (number of 0.662 MeV photons per decay)

$h\nu$=0.662 MeV (photon energy)

ρ_{air}=1.293$\frac{101.3·273.1}{101.3·293.1}$=1.205 kg m^{-3} ($T$=293.1 K, p=101.3 kPa)

$(\mu/\rho)_{air}$ = 0.007752 m^2 kg^{-1} (mass attenuation coefficient for air)

$(\mu_{tr}/\rho)_{air}$=0.00294 m^2 kg^{-1} (mass energy transfer coefficient for air)

h=1.5 m (height of calculation point above ground)

$(\mu h)_{air}$ = 0.007752 · 1.5 · 1.205

θ=89.57 (opening angle)

$\sec\beta$=133.3

Data inserted in Eq. (6.3.87) gives

$$\dot{\Phi} = \frac{80000 \cdot 0.898 \cdot 0.946}{2}[E_1(7.752 \cdot 10^{-3} \cdot 1.5 \cdot 1.205)-E_1(7.752 \cdot 10^{-3} \cdot 1.5 \cdot 1.205 \cdot 133.3)]$$

$$\dot{\Phi} = 33980[E_1(1.395 \cdot 10^{-2}) - E_1(1.868)] \text{ m}^{-2} \text{ s}^{-1}$$

E_1 is obtained from Table 6.5. This gives

$E_1(0.01395)$=3.721 and $E_1(1.868)$=0.059

This inserted gives $\dot{\Phi} = 33980(3.721 - 0.059)$ m^{-2} s^{-1}

and

$$\dot{K} = 33980(3.721 - 0.059) \cdot 0.662 \cdot 1.602 \cdot 10^{-13} \cdot 0.00294 = 38.8 \cdot 10^{-12}$$
Gy s^{-1} =0.140 μGy h^{-1}

Answer: The maximal air kerma rate is 0.14 μGy h^{-1}.

Solution exercise 6.24.

The fluence rate, due to primary photons centrally above the circular area with infinite radius covered with snow, is given by the equation

$$\dot{\Phi}_{P,A} = \frac{S_A f}{2}E_1(\mu_{snow}x_{snow} + \mu_{air}x_{air}) \qquad (6.3.89)$$

where E_1 is the exponential integral of first order.

 To calculate the fluence rate from secondary photons the build-up factor given by the Berger expression

$$B = 1 + a\mu x e^{b\mu x} \qquad (6.3.90)$$

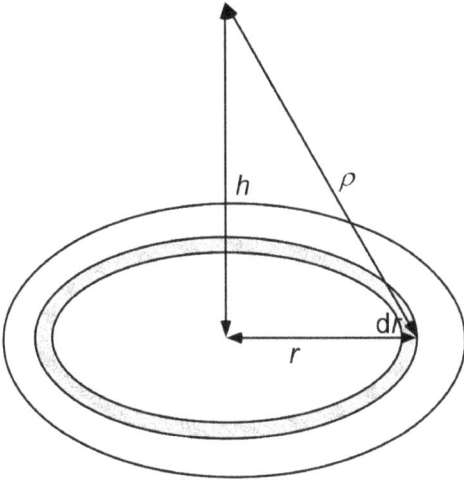

Figure 6.17: Circular surface area source for exercise 6.24.

is used. The second factor gives the ratio of the secondary to the primary photons. Thus the contribution from the secondary photons is given by

$$\dot{\Phi}_{S,A} = S_A f \int_0^\infty \frac{a\mu x \rho e^{-(1-b)\mu x (\rho/h)} 2\pi r \, dr}{4\pi\rho^2 h} \tag{6.3.91}$$

where μx is total attenuation thickness and r is the radius and $\rho = \sqrt{h^2 + r^2}$ (see Fig. 6.17).

Substituting r for ρ and rearranging, the integral can be rewritten as

$$\dot{\Phi}_{S,A} = \frac{S_A f a 2\pi\mu \, x}{4\pi h} \int_h^\infty \frac{\rho\rho e^{-(1-b)\mu x (\rho/h)}}{\rho^2} \, d\rho \tag{6.3.92}$$

Solving the integral gives

$$\dot{\Phi}_{S,A} = \frac{S_A f a}{2(1-b)} e^{-(1-b)\mu x} \tag{6.3.93}$$

The corresponding kerma rate in air for a fluence rate $\dot{\Phi}$ is given by

$$\dot{K} = \dot{\Phi} \sum h\nu_i (\mu_{tr}/\rho)_{i,air} \tag{6.3.94}$$

In principle there should be different kerma factors for primary and secondary photons as they have different energies. However, assuming that the Berger expression holds for kerma or dose, the same kerma factor will be applied for primary and secondary photons.

Data:

S_A = 10000 Bq m^{-2} (area source strength)

f=0.80 (number of photons per decay)

hν=0.80 MeV (photon energy)

$(\mu_{tr}/\rho)_{air}$ = 0.00289 m^2 kg^{-1} (mass energy transfer coefficient for air)

ρ_{snow} = 0.25 · 10^3 kg m^{-3} (density for snow)

μ_{snow} = 0.00787 · 0.25 · 10^3 m^{-1} (linear attenuation coefficient for snow)

ρ_{air} = 1.297 kg m^{-3} (density for air if T=268 K, p=100 kPa)

μ_{air} = 0.00707 · 1.297 m^{-1} (linear attenuation coefficient for air)

x_{snow}=50 cm (snow thickness)

x_{air}=100 cm (thickness of air)

a=1.74, b=0.045 (obtained for Berger parameters for water)

Kerma rate without snow

Data inserted gives:

Primary photons

$E_1(1.50 \cdot 0.00707 \cdot 1.297) = E_1(1.375 \cdot 10^{-2})$

Table 6.5 gives E_1=3.74

The primary photon fluence rate is then given by Eq. 6.3.89

$\dot{\Phi}_P$ = 10000 · 0.8 · 3.74/2 =1.496·10^4 m^{-2} s^{-1}

Secondary photons

Eq. 6.3.93 gives

$$\dot{\Phi}_S = \frac{10000 \cdot 0.8 \cdot 1.74}{2(1 - 0.045)} e^{-(1-0.045)1.375 \cdot 10^{-2}} = 7.19 \cdot 10^3 \text{m}^{-2} \text{s}^{-1}$$

Total photon fluence rate is then

$\dot{\Phi}$ = 1.496 · 10^4 + 7.19 · 10^3 = 2.215 · 10^4 m^{-2} s^{-1}

The air kerma rate is

$\dot{K}$ = 2.215 · 10^4 · 0.00289 · 0.8 · 1.602 · 10^{-13}=8.204·10^{-12} Gy s^{-1}

$\dot{K}$ =0.26 mSv (year)$^{-1}$

Kerma rate with snow

Data inserted gives

Primary photons
$$E_1(100 \cdot 0.0707 \cdot 1.297 \cdot 10^{-3} + 0.0787 \cdot 0.25 \cdot 50) = E_1(0.993)$$

Table 6.5 gives $E_1 = 0.220$

The primary photon fluence rate is then given by

$$\dot{\Phi}_{P,A} = 10000 \cdot 0.8 \cdot 0.220/2 = 880 \text{ m}^{-2} \text{ s}^{-1}$$

Secondary photons
$$\dot{\Phi}_{S,A} = \frac{10000 \cdot 0.8 \cdot 1.74}{2(1 - 0.045)} e^{-(1-0.045)0.993} = 2.823 \cdot 10^3 \text{ m}^{-2} \text{ s}^{-1}$$

Total photon fluence rate is then

$$\dot{\Phi}_A = 880 + 2.823 \cdot 10^3 = 3.703 \cdot 10^3 \text{ m}^{-2} \text{ s}^{-1}$$

The air kerma rate is

$$\dot{K} = 3.703 \cdot 10^3 \cdot 0.00289 \cdot 0.8 \cdot 1.602 \cdot 10^{-13} = 1.371 \cdot 10^{-12} \text{ Gy s}^{-1}$$

$$\dot{K} = 0.043 \text{ mGy (year)}^{-1}$$

Answer: The air kerma rate without snow is $0.26 \text{ mGy (year)}^{-1}$ and with snow $0.043 \text{ mGy (year)}^{-1}$

Solution exercise 6.25.
The primary photon fluence rate at P is given by (see Fig. 6.18)

$$\dot{\Phi}_P = \int_{H}^{H+L} \frac{S_L f e^{-\mu_{\text{water}} H} e^{-\mu_{\text{rod}}(x-H)}}{4\pi x^2} \, dx \tag{6.3.95}$$

where μ_{rod} is the linear attenuation coefficient in the uranium rod, μ_{water} is the linear attenuation coefficient in water, and S_L is linear source strength and f is the number of photons per decay.

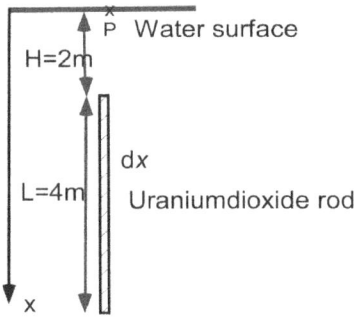

Figure 6.18: Uraniumdioxide rod placed in a water basin.

Rewriting the equation gives

$$\dot{\Phi}_P = \frac{S_L f e^{-H(\mu_{\text{water}} - \mu_{\text{rod}})}}{4\pi} \int_{H}^{H+L} \frac{e^{-\mu_{\text{rod}} x}}{x^2} dx \qquad (6.3.96)$$

Substitute $\mu_{\text{rod}} x = y$, $dx = dy / \mu_{\text{rod}}$

$$\dot{\Phi}_P = \frac{S_L f e^{-H(\mu_{\text{water}} - \mu_{\text{rod}})}}{4\pi} \int_{\mu_{\text{rod}} H}^{\mu_{\text{rod}}(H+L)} \frac{\mu_{\text{rod}} e^{-y}}{y^2} dy \qquad (6.3.97)$$

Using the definition of exponential integrals the solution is

$$\dot{\Phi}_P = \frac{S_L f e^{-H(\mu_{\text{water}} - \mu_{\text{rod}})}}{4\pi} \left[\frac{E_2(\mu_{\text{rod}} H)}{H} - \frac{E_2(\mu_{\text{rod}}(H+L))}{H+L} \right] \qquad (6.3.98)$$

For large z, $E_2(z) \approx \frac{e^{-z}}{z}$ and thus $E_2(\mu_{\text{rod}} H) \approx \frac{e^{-\mu_{\text{rod}} H}}{\mu_{\text{rod}} H}$

Also, $\frac{E_2(\mu_{\text{rod}} H)}{H}$ is much larger than $\frac{E_2(\mu_{\text{rod}}(H+L))}{H+L}$. This means that the second term may be neglected.

This gives

$$\dot{\Phi}_P = \frac{S_L f e^{-H(\mu_{\text{water}} - \mu_{\text{rod}})}}{4\pi} \frac{e^{-\mu_{\text{rod}} H}}{\mu_{\text{rod}} H H} = \frac{S_L f e^{-H \mu_{\text{water}}}}{4\pi \mu_{\text{rod}} H^2} \qquad (6.3.99)$$

Data:
To calculate $(\mu/\rho)_{\text{rod}}$ for uraniumdioxide (UO_2) use the Bragg additivity rule.

$$(\mu/\rho)_{UO_2} = \omega_U (\mu/\rho)_U + \omega_O (\mu/\rho)_O \qquad (6.3.100)$$

Mass fraction: $\omega_U : 238.2/270.3 = 0.8816$; $\omega_O : 32/270.3 = 0.1184$

$(\mu/\rho)_U = 0.00754\,\mathrm{m^2\,kg^{-1}}$, $(\mu/\rho)_O = 0.00637\,\mathrm{m^2\,kg^{-1}}$
$(\mu/\rho)_{rod} = 0.8816 \cdot 0.00754 + 0.1184 \cdot 0.00637 = 0.00740\,\mathrm{m^2\,kg^{-1}}$
$\rho_{UO_2} = 10.96 \cdot 10^3\,\mathrm{kg\,m^{-3}}$ (density of uraniumdioxide)
$(\mu/\rho)_{water} = 0.00707\,\mathrm{m^2\,kg^{-1}}$ (mass attenuation coefficient for water)
$\rho_{water} = 1.00 \cdot 10^3\,\mathrm{kg\,m^{-3}}$ (density of water)
$S_L = 3.7 \cdot 10^{16}/4\,\mathrm{Bq\,m^{-1}}$ (line source strength)
$f = 1.0$ (number of photons per decay)
$H = 2.0\,\mathrm{m}$ (distance from line source to water surface)
$L = 4.0\,\mathrm{m}$ (length of line source)

Data inserted gives

$$\dot{\Phi}_P = \frac{3.7 \cdot 10^{16} \cdot 1 \cdot e^{-2 \cdot 0.00707 \cdot 1000}}{4 \cdot 4\pi \cdot 0.00740 \cdot 10.96 \cdot 10^3 \cdot 2^2} = 1.640 \cdot 10^6\,\mathrm{m^{-2}s^{-1}}$$

This solution however probably underestimates the fluence rate at the water surface. This fluence rate is obtained only just above the line source and photons emitted nearly parallel with the source will only be absorbed in water. An estimate of the maximum fluence rate may be obtained if the attenuation coefficient for uranium dioxide is exchanged for the attenuation coefficient for water. Then the relation will be

$$\dot{\Phi}_P = \frac{S_L \cdot e^{-0}}{4\pi}[\frac{E_2(\mu_{water}(2))}{2} - \frac{E_2(\mu_{water}(6))}{6}] \qquad (6.3.101)$$

This will give the fluence rate

$\dot{\Phi}_P = 6.63 \cdot 10^7\,\mathrm{m^{-2}\,s^{-1}}$

This is 40 times larger than when using the self absorption in the uranium dioxide rod.

Answer: The fluence rate directly above the uranium rod is $1.64 \cdot 10^6\,\mathrm{m^{-2}\,s^{-1}}$. However close to this point the fluence rate is probably much higher due to less attenuation in water, compared to in uranium dioxide.

Solution exercise 6.26.
1) Primary photons. Infinite extension of the wall.

a) Without Pb

According to the equation for an infinite volume source, the fluence rate from primary photons at the surface is given by

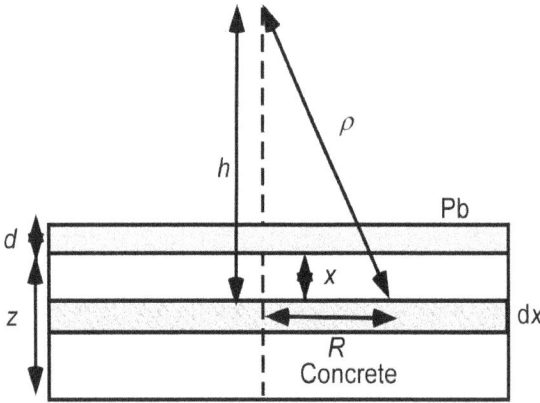

Figure 6.19: Radioactive concrete wall with lead shielding.

$$\dot{\Phi}_{P,V} = \frac{S_V f}{2\mu_s}[1 - E_2(\mu_s z)] \tag{6.3.102}$$

Data:
$\mu_s = 6.495 \cdot 10^{-3} \cdot 2.34 \cdot 10^3 = 15.20\,\text{m}^{-1}$ (linear attenuation coefficient for the concrete wall)
$z = 0.40$ m (wall thickness)
$S_V = 7.4 \cdot 10^3\,\text{Bq}\,\text{m}^{-3}$ (volume source strength)
$f = 2$ (number of photons per decay)

Data inserted in Eq. (6.3.102) gives (see Table 6.5)

$$\dot{\Phi}_{P,V} = \frac{2 \cdot 7.4 \cdot 10^3}{2 \cdot 15.20}[1 - E_2(15.20 \cdot 0.4)] = 4.868 \cdot 10^2(1 - 2.91 \cdot 10^{-4}) = 487\,\text{m}^{-2}\text{s}^{-1}$$

$\dot{\Phi}_{P,V} = 487\,\text{m}^{-2}\,\text{s}^{-1}$

b) With Pb

With extra shielding material the equation for the fluence rate of primary photons becomes

$$\dot{\Phi}_{P,V} = \frac{S_V f}{2\mu_s}[E_2(\mu_{Pb}d) - E_2(\mu_{Pb}d + \mu_s z)] \tag{6.3.103}$$

Data:
$\mu_{Pb} = 7.102 \cdot 10^{-3} \cdot 11.35 \cdot 10^3 = 80.54\,\text{m}^{-1}$ (linear attenuation coefficient for Pb)
$d = 0.020$ m (Pb thickness)

Data inserted in Eq. 6.3.103 gives

$$\dot{\Phi}_{P,V} = \frac{2 \cdot 7.4 \cdot 10^3}{2 \cdot 15.20} [E_2(80.54 \cdot 0.02) - E_2(80.54 \cdot 0.02 + 15.20 \cdot 0.4)] \, \text{m}^{-2}\text{s}^{-1}$$

$$\dot{\Phi}_{P,V} = 4.87 \cdot 10^2 [E_2(1.611) - E_2(7.69)] = 4.87 \cdot 10^2 [0.064 - 5.51 \cdot 10^{-5}] = 31 \, \text{m}^{-2}\text{s}^{-1}$$

2) Secondary photons.

Use the expression by Berger for the build-up factor.

$$B = 1 + a\mu x e^{b\mu x}$$

For a plane source with radius R the fluence rate from secondary photons at a height h is without external attenuation then given by (see Fig. 6.19).

$$\dot{\Phi}_S = S_A f \int_0^R \frac{a\mu_s x \rho e^{-(1-b)\mu_s x \frac{\rho}{h}} 2\pi r \, dr}{h 4\pi\rho^2} = \frac{S_A f}{2} \int_h^{\sqrt{h^2+R^2}} \frac{a\mu_s x \rho\rho e^{-(1-b)\mu_s x \frac{\rho}{h}} \, d\rho}{h\rho^2}$$

The equation is reduced to

$$\dot{\Phi}_{S,A} = \frac{S_A f}{2} \int_h^{\sqrt{h^2+R^2}} \frac{a\mu_s x e^{-(1-b)\mu_s x \frac{\rho}{h}} \, d\rho}{h} \tag{6.3.104}$$

Substitute $y = \frac{\mu_s x \rho}{h}$; $dy = \frac{\mu_s x \, d\rho}{h}$

$$\dot{\Phi}_{S,A} = \frac{S_A f a}{2} \int_{\mu_s x}^{\mu_s x \sqrt{1+\frac{R^2}{h^2}}} \frac{h\mu_s x e^{-(1-b)y}}{\mu_s x h} \, dy = \frac{S_A f a}{2} \int_{\mu_s x}^{\mu_s x \sqrt{1+\frac{R^2}{h^2}}} e^{-(1-b)y} \, dy \tag{6.3.105}$$

$$\dot{\Phi}_{S,A} = \frac{S_A f a}{2(b-1)} [e^{-(1-b)\mu_s x \sqrt{1+\frac{R^2}{h^2}}} - e^{-(1-b)\mu_s x}] \tag{6.3.106}$$

Assume infinite extension of the wall. Then $R \to \infty$ and

$$\dot{\Phi}_S = \frac{S_A f a}{2(1-b)} e^{-(1-b)\mu_s x} \tag{6.3.107}$$

The plane source is integrated over the wall thickness z and including the Pb-shield, the fluence rate will be given by

$$\dot{\Phi}_{S,V} = \frac{S_V f a}{2(1-b)} \int_0^z e^{-(1-b)(\mu_{Pb} d + \mu_s x)} \, dx = \frac{S_V f a e^{-(1-b)\mu_{Pb} d}}{2(1-b)} \int_0^z e^{-(1-b)\mu_s x} \, dx \tag{6.3.108}$$

Solving the integral gives

$$\dot{\Phi}_{S,V} = \frac{S_V fa e^{-(1-b)\mu_{Pb}d}}{2(1-b)^2 \mu_s}[1 - e^{-(1-b)\mu_s z}]$$

(6.3.109)

a) Without Pb

Then $d=0$ and Eq. 6.3.109 is reduced to

$$\dot{\Phi}_{S,V} = \frac{S_V fa}{2(1-b)^2 \mu_s}[1 - e^{-(1-b)\mu_s z}]$$

(6.3.110)

Data:

a=1.27, b=0.032 (parameters for Berger equation for concrete)

Data inserted in Eq. (6.3.110) gives

$$\dot{\Phi}_{S,V} = \frac{2 \cdot 7.4 \cdot 10^3 \cdot 1.27}{2(1-0.032)^2 15.20}[1 - e^{-(1-0.032)15.20 \cdot 0.4}] = 658\,\text{m}^{-2}\text{s}^{-1}$$

$\dot{\Phi}_{S,V}=658\,\text{m}^{-2}\,\text{s}^{-1}$

b) With Pb

In the Berger expression use data for lead over the whole thickness as lead is the outermost material ($a=0.30$, $b=-0.015$).

Data inserted in Eq. (6.3.109) gives

$$\dot{\Phi}_{S,V} = \frac{2 \cdot 7.4 \cdot 10^3 \cdot 0.30 \cdot e^{-(1+0.015)0.02 \cdot 80.51}}{2(1+0.015)^2 \cdot 15.20}[1 - e^{-(1+0.015)15.20 \cdot 0.4}]$$

$\dot{\Phi}_{S,V}=27.6\,\text{m}^{-2}\,\text{s}^{-1}$

This probably underestimates the contribution. If instead the Berger parameters for concrete are used the secondary fluence rate is 139 m^{-2} s^{-1}. This is then instead an overestimation.

Total fluence rate is thus

a) Without Pb: 487+658=1145 m^{-2} s^{-1}

b) With Pb: 31+28=59 m^{-2} s^{-1}

Discussion of the approximation of infinite wall area.

Consider only primary photons.

I. Infinite area source
$$\Phi_\infty = (S_A/2)E_1(\mu_s x)$$

II. Finite area source
$$\dot{\Phi}_{area} = (S_A/2)[E_1(\mu_s x) - E_1(\mu_s x(\rho/h))]$$

where x is the thickness of the absorber of lead, h is the height from area source to calculation point, and ρ is the distance from calculation point to outer radius of the source.

Assume that the wall may be approximated with a circular area equal to the real area ($6.0 \times 2.0 = 12.0 \, m^2$). Then the radius will be R=1.95 m. This gives $\rho/h = \sqrt{1 + 1.95^2}/1 = 2.19$ and $\mu_{Pb}x \cdot \rho/h = 3.53$

This gives the fluence rates

$$\dot{\Phi}_\infty = (S_A/2)E_1(1.61) = (S_A/2) \cdot 8.57 \cdot 10^{-2} \, m^{-2} \, s^{-1}$$

$$\dot{\Phi}_{area} = (S_A/2)(8.57 \cdot 10^{-2} - 6.78 \cdot 10^{-3}) = (S_A/2) \cdot 7.89 \cdot 10^{-2} \, m^{-2} \, s^{-1}$$

The ratio between the two fluences is then 1.08. Assuming a similar relation for the secondary photons the approximation of an infinite area then overestimates the fluence with around 10%.

Answer: Assuming infinite extension the fluence rate is without Pb $1.01 \cdot 10^3 \, m^{-2} \, s^{-1}$ and with Pb $1.8 \cdot 10^2 \, m^{-2} \, s^{-1}$. The assumption of infinite extension overestimates the fluence rate with about 10%.

Bibliography

Berger M. J. (1956). Effects of Boundaries and Inhomogeneities on the Penetration of Gamma Radiation. Report NBS-492. Washington,DC, U S National Bureau of Standards.

Berger M. J. and Doggett J. (1956). Reflection and Transmission of Gamma Radiation by Barriers: Semianalytic Monte Carlo Calculation. Journal of Research of the National Bureau of Standards, 56,(2), 89-98.

Chilton A. B. (1968). Buildup Factor. In R. G. Jaeger (Ed.), Engineering Compendium on Radiation Shielding (Sec. 4.3.1.2). New York, NY. Springer-Verlag.

Puck T. T. and Markus P. L. (1956). Action of x-rays on mammalian cells. Journal of Experimental Medicine, 103, 653-666.

Table 6.4: Data for the Sievert integral (Sievert, 1921) for various values of the parameters $\Theta(5° - 90°)$ and $\Sigma\mu_i x_i$ (0-10).

$\Sigma\mu_i x_i/\Theta$	5°	10°	15°	20°	30°	40°
0.00	0.08727	0.17453	0.26180	0.34906	0.52359	0.69812
0.05	0.08300	0.16597	0.24888	0.33169	0.49684	0.66102
0.10	0.07895	0.15784	0.23661	0.31519	0.47261	0.62588
0.20	0.07143	0.14275	0.21384	0.28445	0.42452	0.56116
0.40	0.05847	0.11675	0.17467	0.23204	0.34421	0.45120
0.60	0.04786	0.09549	0.14268	0.18919	0.27912	0.36289
0.80	0.03917	0.07826	0.11655	0.15426	0.22635	0.29196
1.00	0.03206	0.06388	0.09520	0.12577	0.18358	0.23496
1.25	0.02496	0.04969	0.07393	0.09745	0.14131	0.17916
1.50	0.01943	0.03865	0.05741	0.07551	0.10878	0.13668
1.75	0.01513	0.03006	0.04458	0.05851	0.08375	0.10431
2.00	0.01178	0.02357	0.03462	0.04533	0.06449	0.07964
2.50	0.00714	0.01431	0.02088	0.02722	0.03825	0.04649
3.00	0.00433	0.00856	0.01259	0.01635	0.02270	0.02718
3.50	0.00262	0.00518	0.00760	0.00982	0.01348	0.01591
4.00	0.00159	0.00313	0.00458	0.00590	0.00800	0.00933
5.00	0.00058	0.00115	0.00167	0.00213	0.00283	0.00322
6.00	0.00021	0.00042	0.00061	0.00077	0.00100	0.00112
8.00	0.00003	0.00006	0.00008	0.00010	0.00013	0.00014
10.00	0.0000	0.00001	0.00001	0.00001	0.00002	0.00002

$\Sigma\mu_i x_i/\Theta$	50°	60°	70°	80°	90°
0.00	0.87265	1.04718	1.22171	1.39624	1.57077
0.05	0.82359	0.98346	1.13830	1.28119	1.36517
0.10	0.77725	0.92378	1.06115	1.17832	1.22863
0.20	0.69256	0.81547	0.92368	1.00282	1.02368
0.40	0.55015	0.63677	0.70403	0.74075	0.74521
0.60	0.43743	0.49850	0.54043	0.55782	0.55889
0.80	0.348 11	0.39120	0.38776	0.42578	0.42797
1.00	0.27727	0.30768	0.32411	0.32867	0.32829
1.25	0.20865	0.22858	0.23888	0.23948	0.23949
1.50	0.15755	0.17033	0.17548	0.17621	0.17621
1.75	0.11898	0.12727	0.13018	0.13049	0.13049
2.00	0.09049	0.09534	0.09699	0.09712	0.09712
2.50	0.05159	0.05387	0.05440	0.05442	0.05442
3.00	0.02970	0.03067	0.03084	0.03085	0.03085
3.50	0.01716	0.01758	0.01763	0.01763	0.01763
4.00	0.00995	0.01013	0.01015	0.01015	0.01015
5.00	0.00337	0.00341	0.00341	0.00341	0.00341
6.00	0.00116	0.00116	0.00116	0.00116	0.00116
8.00	0.00014	0.00014	0.00014	0.00014	0.00014
10.00	0.00017	0.00017	0.00017	0.00017	0.00017

Table 6.5: Exponential integral functions, $E_1(x)$ and $E_2(x)$. for x between 0 and 10. Data taken from Handbook of Mathematical Functions (Abramowitz and Stegun Eds, 1964)

x	$E_1(x)$	$E_2(x)$
0.0	∞	1.0000
0.01	4.03793	0.94967
0.015	3.63743	0.93053
0.02	3.35471	0.91310
0.025	3.31365	0.89688
0.03	2.95912	0.88166
0.04	2.68127	0.85353
0.05	2.46790	0.82783
0.06	2.29531	0.80405
0.07	2.15084	0.78183
0.08	2.02695	0.76095
0.09	1.91875	0.74124
0.10	1.82292	0.72254
0.15	1.46447	0.64103
0.20	1.22265	0.57419
0.25	1.04428	0.51773
0.30	0.90568	0.46911
0.35	0.79419	0.42671
0.40	0.70238	0.38936
0.45	0.62533	0.35622
0.50	0.55977	0.32664
0.60	0.45437	0.27618
0.70	0.37376	0.23494
0.80	0.31059	0.20085
0.90	0.26018	0.17240
1.00	0.21938	0.14849
1.10	0.18599	0.12828
1.20	0.15840	0.11110
1.30	0.13545	0.09644
1.40	0.11621	0.08388
1.50	0.10001	0.07310
1.60	0.08630	0.06380
1.70	0.07465	0.05577
1.80	0.06471	0.04881
2.00	0.04890	0.03753
2.50	0.02492	0.01980
3.00	0.01305	0.01064
3.50	0.00697	0.00580
4.00	0.00378	0.00320
5.00	0.00115	0.00010
6.00	0.00036	0.00032
7.00	1.15E-4	1.04E-4
8.00	3.77E-5	3.41E-5
9.00	1.24E-5	1.14E-5
10.00	4.16E-6	3.83E-6

Bibliography

References A. Tables of physical constants and radioactive decay

Berger M. J., Hubbell J. H., Seltzer S. M., et al. (2005). XCOM: Photon Cross Section Database (version 1.3). Gaithersburg, MD:National Institute of Standards and Technology. Available: http://physics.nist.gov/xcom [2009, October].

Berger M. J., Coursey J. S., Sucker D. S., et al. (2005). ESTAR, PSTAR and ASTAR. Computer Programs for Calculating Stopping-Power Tables for Electrons,Protons and Helium Ions.[Online]. . Gaithersburg, MD:National Institute of Standards and Technology. Available: http://physics.nist.gov/Star. [2013, Oct].

Browne E., and Firestone R. B. (1986). Table of Radioactive Isotopes. New York,NY:Wiley.

Eckerman K. F., Endo A. (2008), MIRD: Radionuclide Data and Decay Schemes. New York, NY:The Society of Nuclear Medicine.

ICRP. (1983). Radionuclide transformations. Energy and Intensity of Emissions. ICRP Publication 38. Oxford-New York-Frankfurt. Pergamon Press.

ICRP. (2008). Nuclear Decay Data for Dosimetric Calculations. ICRP Publication 107. Annals of ICRP 38 (3).

International Commission on Radiological Units and Measurements.(2011). Fundamental Quantities and Units for Ionizing Radiation. ICRU Report 85. Bethesda, MD:ICRU Publications.

Lederer C. M. and Shirley V. S. (1978). Table of Isotopes. 7th ed. New York, NY:Wiley.

Nordling C. and Osterman J. (2006). Physics Handbook for science and engineering. Lund, Sweden:Studentlitteratur AB.

References   B. General background information

Anderson D. W. Absorption of Ionizing Radiation. (1994). Baltimore, MD:University Park Press.

Attix F. H. (1986). Introduction to Radiological Physics and Radiation Dosimetry.New York,NY:Wiley.

Chilton A. B., Shultis J. K. and Faw R. E. (1984). Principles of Radiation Shielding.Englewood Cliffs, NJ:Prentice-Hall Inc.

Evans R. D. (1955) The Atomic Nucleus. New York,NY:McGraw-Hill.

Hall E. J. and Giaccia, A. M. (2005). Radiobiology for the Radiologist. Hagerstown, MD:Lippincott Williams and Wilkins.

ICRP. (2007). The 2007 Recommendations of the International Commission on Radiological Protection. ICRP Publication 103, Annals of ICRP 37 (2-4).

Knoll G. F. (2010). Radiation Detection and Measurement. Fourth ed. New York, NY, Wiley.

Krane K. (1987). Introductory Nuclear Physics. New York, NY, Wiley.

Podgorsak E. B. (2010). Radiation Physics for Medical Physicists. Berlin, Germany:Springer.

Stabin M.G. (2007). Radiation Protection and Dosimetry: An Introduction to Health Physics. New York, NY:Springer.

Index

Erratum

ISBN (Online): 9783110442069

DOI (Chapter): https://doi.org/10.1515/9783110442069-001

https://doi.org/10.1515/9783110442069-002

https://doi.org/10.1515/9783110442069-004

https://doi.org/10.1515/9783110442069-006

DOI (Book): https://doi.org/10.1515/9783110442069

DOI of the Erratum: https://doi.org/10.1515/9783110442069-009

1. Solution exercise 1.12

Eq. 1.3.47 now reads:

$$N_{2,\mathrm{Sm}} = \int_0^T A_{\mathrm{Pm}}\, \mathrm{d}t = \frac{\lambda_{\mathrm{Pm}} A_{\mathrm{Nd}}}{\lambda_{\mathrm{Nd}} - \lambda_{\mathrm{Pm}}} \left[-\frac{e^{\lambda_{\mathrm{Pm}}t}}{\lambda_{\mathrm{Pm}}} + \frac{e^{\lambda_{\mathrm{Nd}}t}}{\lambda_{\mathrm{Nd}}} \right]_0^T$$

Should read

$$N_{2,\mathrm{Sm}} = \int_0^T A_{\mathrm{Pm}}\, \mathrm{d}t = \frac{\lambda_{\mathrm{Pm}} A_{\mathrm{Nd}}}{\lambda_{\mathrm{Nd}} - \lambda_{\mathrm{Pm}}} \left[-\frac{e^{-\lambda_{\mathrm{Pm}}t}}{\lambda_{\mathrm{Pm}}} + \frac{e^{-\lambda_{\mathrm{Nd}}t}}{\lambda_{\mathrm{Nd}}} \right]_0^T$$

2. Exercise 2.16

Now reads:

Calculate the energy of a Compton scattered photon with the primary energy 1.17 MeV if it is scattered in the angle $\pi/2$ radians a) in one scattering, b) in two equal scatterings c) in three equal scatterings d) in an infinite number of equal small scatterings.

Should read :

Calculate the energy of a Compton scattered photon with the primary energy 1.17 MeV if it is scattered in the angle π radians a) in one scattering, b) in two equal scatterings c) in three equal scatterings d) in an infinite number of equal small scatterings.

3. Solution exercise 4.16.

Heading now reads:

Solution 4.16

Should read:

Solution exercise 4.16

Last line and answer in this solution reads:

The ratio of the absorbed doses with and without the polystyrene cover is 1.017.

Answer: The ratio of the absorbed doses in the diamond detector with and without the polystyrene cover is 1.017.

Should read:

The ratio of the absorbed doses with and without the polystyrene cover is 0.984.

Answer: The ratio of the absorbed doses in the diamond detector with and without the polystyrene cover is 0.984.

4. Solution exercise 6. 4.

The solution reads:

The variation of ^{210}Po activity in the body is given by the retention function corrected with the physical half-life.

$$A(t) = A_0 e^{-\lambda_{\mathrm{eff}} t} \tag{6.3.11}$$

$\lambda_{\mathrm{eff}} = \lambda_{\mathrm{f}} + \lambda_{\mathrm{b}}$

The excretion is given by differentiation of Eq. (6.3.11)

$$\frac{\mathrm{d}A}{\mathrm{d}t} = -A_0 \lambda_{\mathrm{eff}} e^{-\lambda_{\mathrm{eff}} t} \tag{6.3.12}$$

where
A_0=20.0 kBq (initial activity of ^{210}Po)
λ_{f}=ln2/138.4 d^{-1} (physical decay constant)
λ_{b}=ln2/40 d^{-1} (biological decay constant)
λ_{eff}=2.234·10^{-2} d^{-1} (effective decay constant)

The minimum measured activity excreted in the daily urine should be 15 Bq. 10% of the activity is excreted through the urine. Thus data inserted Eq. (6.3.12) gives

$$15 = 0.1 \cdot 20 \cdot 10^3 \cdot 2.234 \cdot 10^{-2} \cdot e^{-2.234 \cdot 10^{-2} t}$$

Solving t gives f=49d.

Answer: It is necessary to measure at least every 49th day.

Should read:

The variation of ^{210}Po activity in the body is given by the retention function corrected with the physical half-life.

$$A(t) = A_0 e^{-\lambda_{\text{eff}} t} \qquad\qquad (6.3.11)$$

$\lambda_{\text{eff}} = \lambda_f + \lambda_b$

As the decrease due to the release of the activity from the body is given by the biological decay constant, λ_b, the excretion is given by

$$\frac{dA}{dt} = -A_0 \lambda_b e^{-\lambda_{\text{eff}} t} \qquad\qquad (6.3.12)$$

where
A_0=20.0 kBq (initial activity of ^{210}Po)
λ_f=ln2/138.4 d^{-1} (physical decay constant)
λ_b=ln2/40 d^{-1} (biological decay constant)
λ_{eff}=2.234·10^{-2} d^{-1} (effective decay constant)

The minimum measured activity excreted in the daily urine should be 15 Bq. 10% of the activity is excreted through the urine. Thus data inserted Eq. (6.3.12) gives

$$15 = 0.1 \cdot 20 \cdot 10^3 (\ln2/40) \cdot e^{-2.234\cdot10^{-2}t}$$

Solving t gives f=37.4 d.

Answer: It is necessary to measure at least every 37th day.

www.ingramcontent.com/pod-product-compliance
Lightning Source LLC
Chambersburg PA
CBHW040138200326
41458CB00025B/6310

* 9 7 8 3 1 1 0 4 4 2 0 5 2 *